RANCHO SA
OR MAIN HD9651.
Du Pont and the
Taylor, Graham D.
P9-DTG-422
3 3065 00032 5377

DU PONT AND
THE INTERNATIONAL
CHEMICAL INDUSTRY

Executive Committee of E. I. duPont de Nemours & Co., Inc., 1919. Standing (left to right): J. J. Raskob, R. R. M. Carpenter, W. M. Coyne, F. G. Tallman. Seated (left to right): F. L. Connable, H. G. Haskell, Irénée du Pont, H. F. Brown, Lammot du Pont. *Courtesy of the Eleutherian Mills Historical Library.*

THE EVOLUTION OF AMERICAN BUSINESS: INDUSTRIES, INSTITUTIONS, AND ENTREPRENEURS

Albro Martin, Series Editor
Oglesby Professor of the American Heritage
Bradley University

Editorial Advisory Board
Alfred D. Chandler, Jr., Isidore Strauss Professor of Business History,
Harvard Graduate School of Business Administration

Diane Lindstrom, Associate Professor of History and Women's Studies,
University of Wisconsin, Madison

Harold C. Livesay, Professor of History,
Virginia Polytechnic Institute

Peter D. McClelland, Professor of Economics,
Cornell University

GRAHAM D. TAYLOR

PATRICIA E. SUDNIK

Rancho Santiago College
Orange Campus Library
20[illegible]381

DU PONT AND THE INTERNATIONAL CHEMICAL INDUSTRY

THE EVOLUTION OF AMERICAN BUSINESS:
INDUSTRIES, INSTITUTIONS, AND ENTREPRENEURS

Copyright © 1984 by G. K. Hall & Company
All Rights Reserved
Published by Twayne Publishers
A Division of G. K. Hall & Company
70 Lincoln Street
Boston, Massachusetts 02111

Book production by Marne B. Sultz

Book design by Barbara Anderson

Typeset in 10 point Times Roman
with Univers display type
by Compset, Inc., Beverly, Massachusetts

HD
9651.9
.D8
T39
1984

Printed on permanent/durable
acid-free paper and bound in
the United States of America

Library of Congress Cataloging in Publication Data

Taylor, Graham D., 1944-
Du Pont and the international chemical industry.

(The Evolution of American business)
Bibliography: p. 231
Includes index.
1. E.I. du Pont de Nemours & Company—History.
2. Chemical industry—United States—History.
3. Chemical industry—History. I. Sudnik,
Patricia E.
II. Title. III. Series.
HD9651.9.D8T39 1984 338.8′87 84-8933
ISBN 0-8057-9805-6

CONTENTS

PREFACE

The chemical industry is second only to the oil industry in the complexity of its operations and in the ramifications of its influence on modern society. Textiles, fertilizers, preservatives, drugs, detergents, packaging materials, adhesives, machine parts, explosives, refined fuels, rocket propellants, and plastic toys are all products of chemical processes, and this is hardly an exhaustive catalog. Hundreds of other industries, from pulp and paper processors to movie producers, rely on materials supplied by chemical manufacturers, and the relationship between the oil and chemical industries has become increasingly close in the years since World War II.

No one company and no single country dominates this diverse and amorphous industry. For many years France, Belgium, and Britain were the major producers of heavy chemicals. Germany and Switzerland led in the field of organic chemicals. The United States has emerged as the major innovator in the products of polymer chemistry, particularly in synthetic fibers. Today virtually every industrial nation and many developing countries have established some form of domestic chemical manufacturing, although the United States and Western Europe remain the major exporters of most materials.

Chemical technology is complex and more closely aligned to developments in the natural sciences than any other major industry. Few chemical manufacturers can neglect for very long the impact of advances in scientific knowledge on their industry, and for this reason research has been an essential ingredient for survival. Chemical manufacturers pioneered in industrial research and continue today to maintain scientific organizations larger than those in any other industry.

Chemical technology has simultaneously encouraged greater international trade, since its processes require a steady flow of diverse raw materials, and altered patterns of trade by replacing products based on natural sources with synthetics. The cosmopolitan character of chemical technology also contributed to the development of a system of cooperation observed by industry leaders from the middle of the nineteenth century onward. The chemical industry both in the United States and in world markets exhibited an unusual degree of stability during the first fifty years of the twentieth century, a stability based on an intricate structure of "business diplomacy" that performed the tasks of conciliation far more successfully than governments in this period. After World War II this system was disrupted by the abrupt change in the balance of economic power between America and Europe, and by the efforts of the United States government to eliminate "cartels" that were regarded as obstacles to international peace and prosperity. American chemical companies such as Du Pont began to move into export markets and direct investment abroad. But the range of products was so diverse and the tradition of cooperation so ingrained that the gentlemanly observance of spheres of influence remains characteristic of the industry.

The chemical industry is capital-intensive. Not only research costs but also basic plant design and operations contribute to this situation, both for large-scale continuous-production processes in fields such as synthetic fibers and for specialized processes in areas such as dyes. In most product lines, planning must begin almost five years before production, and development costs are substantial. Given the level of capital investment required, the industry's devotion to practices that encourage market stability is not surprising. Related to this concern is the pursuit of economies of scale by chemical manufacturers. Historically, chemical firms have been smaller than the big steel, electrical, and automobile producers in terms of fixed assets, sales volume, and degree of control over markets. But the complexity of chemical processes has compelled manufacturers to devote a great deal of attention to the internal organization of their enterprises. Du Pont in particular was a major innovator in the organization and management of a large corporate structure, and its innovations were adopted by other firms in the chemical industry and beyond.

This study traces the development of Du Pont as part of the international chemical industry and in all of the dimensions outlined here: the ramifications of persistent technological change, the growth of international trade in chemicals, the impact of government policies on the company and the industry, and the significance of innovations in organization and management. Du Pont has been the subject of a variety of studies, ranging from popular exposés to unabashed apologias, and, fortunately for historians, some substantial scholarly analyses, buttressed by more general works on the chemical industry. It has been the centerpiece of numerous investigations and lawsuits that have probed extensively, if not always accurately, into the affairs and operations of the company. The authors of this book have drawn upon these sources, and upon a large body of internal records and interviews with former Du Pont employees. Our aim is to present as thorough and balanced an account as these sources permit of the development of Du Pont, its significance in the evolution of American business, and its role in the growth of one of the most dynamic, controversial, and international industries of the twentieth century.

The authors are particularly indebted to the staff of the Eleutherian Mills Historical Library at Greenville, Delaware, for assistance with the voluminous records of the Du Pont company that are held in its archives, and in particular to Dr. Richmond Williams, director of the library, and Betty Bright-Low, the research and reference librarian. Our thanks are also extended to Dimitri Andriadis of the International Department at Du Pont for arranging interviews with former company employees, and to John Jenney, David Conklin, and Richard Manning for their useful and interesting comments on the company's legal and international operations. It should perhaps be noted that neither the Du Pont company nor family were involved in any way in the funding or preparation of this book. Financial assistance for research was provided by the Social Sciences and Humanities Research Council of Canada, the Dalhousie University Research and Development Fund, and the Centre for International Business Studies at Dalhousie University, with special thanks to Dr. Donald Patton, former director of the Centre.

Our research in the U.S. National Archives benefitted from the help of numerous archivists, in particular David Kepley of the Division of Legislative and Natural Resources Archives, Joseph Howerton and Jerry N. Hess in the Social and Industrial Branch, and Ron Swerczek in the Diplomatic Branch. Access to records of the Antitrust Division of the U.S. Justice Department was provided with the assistance of Leo D. Neshkes, FOI/PA Control Officer, and we were particularly helped by two members of his staff, Robert Huber and James Kasson.

Our thanks must also go to Dr. Albro Martin for helpful editorial comments, Dr. Alfred D. Chandler, Jr. for information on sources, to the editors of Twayne Publishers for their patience and assistance throughout the preparation of the book, and to Mary Wyman for typing several drafts of the manuscript.

This study was very much a "joint venture" in its overall structure and theme, but we did engage in a certain degree of division of labor in writing the book. Graham Taylor was primarily responsible for the prologue and essay on sources, and chapters 1–2, 4–8, 11, and 13–14; and Patricia Sudnik prepared chapters 3 and 9–10. We worked together on chapter 12. Needless to add, we take full responsibility for any errors of omission or commission.

Graham D. Taylor

Halifax, Nova Scotia

Patricia E. Sudnik

Pittsburgh, Pennsylvania

CHRONOLOGY

1791	Development of Leblanc soda ash process (France).
1800	Migration of Du Pont family to the United States.
1802	Establishment of E. I. du Pont de Nemours Company in Delaware.
1834	Death of Eleuthere Irenee du Pont, founder of the powder company.
1845	Development of guncotton by Schönbein (Switzerland).
1856	Development of synthetic dye process by Perkin (Britain).
1857	Development of soda powder by Lammot du Pont I.
1860	Development of ammonia soda process by Solvays (Belgium).
1863	Development of dynamite by Nobel (Sweden).
1872	Formation of "Powder Trust." Establishment of Brunner-Mond Company in ammonia soda field in Britain.
1880	Establishment of Repauno Chemical Company; Du Pont enters dynamite field. Nobel establishes dynamite companies in Britain and Europe.
1894	Du Pont enters smokeless powder field.
1897	Du Pont and Nobel companies establish "international convention" in explosives.

1902 Reorganization of powder company by Du Pont cousins.

1903–1904 Du Pont acquires Laflin and Rand Company; "Powder Trust" dissolved. Consolidation of German dye manufacturers into two cartels.

1907 Antitrust suit brought against Du Pont powder company. Renegotiation of Du Pont-Nobel explosives agreement.

1909 Development of Haber-Bosch process for synthetic ammonia (Germany).

1910 Du Pont enters artificial leather (fabrikoid) field.

1911 Powder company broken up by antitrust decree; formation of Atlas and Hercules Powder Companies. Du Pont and Nobel (U.K.) establish joint venture in Canada: Canadian Explosive Company Ltd.

1914 Du Pont becomes major supplier of powder to Allies. Outbreak of World War I: Du Pont and Nobel agreement suspended.

1915–1918 Du Pont diversifies into paints, plastics, dyes. Christiana Securities Company formed; Pierre du Pont acquires control of Du Pont corporation. Establishment of British Dyestuffs Corporation. German dye cartels combine to form I. G. Farben.

1918–1919 Du Pont investment in General Motors.

1920 Du Pont enters rayon synthetic fiber field. Du Pont and Nobel (U.K.) reestablish patent and process agreement in explosives.

1921 Reorganization of Du Pont; decentralized system adopted.

1922 Protective tariff on dyes and chemicals; establishment of "American Selling Price." B.A.S.F. begins development of synthetic fuel (Germany).

1923 Du Pont acquires cellophane patents.

1924 Du Pont enters synthetic ammonia field.

1925 Reorganization of I. G. Farben as integrated chemical company.

1926 Merger of Nobel (U.K.), Brunner-Mond, British Dyestuffs to form Imperial Chemicals Industries Ltd.

1927–1928 Du Pont invests in U.S. Steel, U.S. Rubber, acquires Grasselli Chemical Company. I. G. Farben and Standard Oil (New Jersey) form joint venture in synthetic fuel, synthetic rubber.

1928–1929 Du Pont and I.C.I. negotiate patent and process agreement ("Grand Alliance").

1934 Nye committee "merchants of death" investigation in Congress.

1935–1937 Development of neoprene and nylon by Carothers. Du Pont and I.C.I. establish joint ventures in South America.

1942 Du Pont participates in Manhattan Project: development of Hanford plutonium plant. German chemical companies in U.S.A. seized by government; I. G. Farben agreement with Standard Oil suspended.

1944 Antitrust suit brought against Du Pont-I.C.I. "Alliance."

1947 Antitrust suit against Du Pont in cellophane field initiated. I. G. Farben executives prosecuted for war crimes at Nuremberg.

1948 Antitrust suit against Du Pont and General Motors initiated. Du Pont and I.C.I. renegotiate patent and process agreement.

1950 Du Pont develops dacron polyester fiber; licenses nylon to Chemstrand Corporation.

1952 Du Pont-I.C.I. agreements and joint ventures broken up by antitrust decree.

1953 Final dissolution of I.G. Farben.

1956 Du Pont wins cellophane antitrust case.

1957 U.S. Supreme Court orders dissolution of Du Pont-General Motors connection. Du Pont initiates direct investment in Europe. Reorganization of Du Pont international operations.

1963–1965 Final dissolution of Du Pont-General Motors connection. German chemical companies, I.C.I. reenter U.S. market.

1981 Du Pont acquires Conoco.

PROLOGUE

On August 4, 1981, the E. I. du Pont de Nemours Company of Wilmington, Delaware acquired 55 percent of the stock of the Continental Oil Company of Stamford, Connecticut. This was the largest merger in American history. Du Pont agreed to pay $7.6 billion for almost 39 million shares in the oil company. The combined assets of the two companies exceeded $18 billion, with combined sales in 1980–81 of $32 billion. Overnight, Du Pont catapulted from the sixteenth to the fifth largest industrial corporation in the United States.

The deal was completed after almost two months of widely publicized maneuvering for control by Du Pont, Seagrams, and Mobil Oil. Seagrams had begun the battle on June 25 when it made what Conoco's management regarded as a hostile bid for 41 percent of the oil company's stock at $73 per share, $12 higher than Conoco's market price. Ralph Bailey, Conoco's chairman, was contacted by his counterpart at Du Pont, the newly installed Edward Jefferson, who offered to act as the "white knight," rescuing Conoco from the clutches of the Bronfmans, Seagrams' Canadian owners. By this time, Conoco's market value had climbed to $75. Du Pont offered $87.50 per share, or an exchange of 1.6 Du Pont common for each share of Conoco. Two weeks after Du Pont's bid Mobil entered the fray initially offering $90 per share, then rapidly escalating to $105 and by the first week of August an astonishing $120 per share for 50 percent of Conoco's stock which was holding at a market price of $90. Despite the allure of Mobil's bid, Conoco's shareholders were restrained by threats from the U.S. Justice department that a Mobil-Conoco merger might produce an antitrust investigation. When Du Pont raised its own bid to $98 and/or 1.7 du Pont shares for Conoco shares, topping Seagrams by $6 per share, the contest was over. In the process, however, Du Pont had not only acquired an oil company but also a not entirely welcome minority shareholder, Seagrams, which emerged with 20 percent of Du Pont stock in exchange for the 17 million-plus shares of Conoco it had picked up during the battle.[1]

Du Pont's takeover of Conoco seemed to be a spur-of-the moment decision, and in certain respects represented a departure from the company's time-honored traditions. In order to swing the deal, Du Pont had to borrow almost $4 billion. In the past it had preferred to draw on its own resources for expansion and the purchase encumbered Du Pont with an unaccustomed debt. Du Pont had no previous direct experience in the oil industry, and after the Conoco takeover some observers expressed doubt that the company could move as rapidly into the biochemicals, especially the promising field of genetic engineering, that Jefferson had earlier proclaimed to be its future source of growth. To this end, Du Pont had proposed to embark on a $4 billion program of development, including an increase in its own research spending for the first time since the 1960s, grants to Harvard University and the California Institute of Technology to support basic research in genetics, and the acquisition of New England Nuclear Company, which produces radioactive isotopes for medical research.[2]

But to Jefferson and other Du Pont executives, the Conoco purchase could be justified in terms their predecessors could appreciate. Despite the move into biochemicals, over one third of the company's sales was in synthetic fibers—nylon, orlon, dacron, and the other materials that Du Pont had developed from the ground up after World War II. An additional 12 percent came from plastics and other organic chemical products. All of these products depended on oil as "feedstock," and the rapid increases in oil prices in 1973 and again in 1979 cut into earnings in these areas. Although oil prices softened in 1980–81, uncertainty about future supplies and costs posed a constant problem. In this context the Conoco acquisition gave Du Pont a secure source of supply that would stabilize costs, and in the longer term would enable the company to apply earnings from these established fields to developing the new biochemical products.

Furthermore, Conoco's huge assets and sales volume—almost 40 percent greater than Du Pont's in 1980–81—would provide Du Pont with a scale of operation and potential resources to meet what Jefferson and his colleagues considered a major challenge of the future: competition from abroad, particularly from the three giant German chemical firms, Bayer, Hoechst, and B.A.S.F. United before World War II as I. G. Farben, the largest chemical organization in the world, the three companies had steadily recovered ground from Du Pont and its British counterpart, Imperial Chemical Industries Ltd., in the years since 1951 when the Allied occupation regime in Germany had sundered the "I. G." By 1979 each of the German "big three" had larger sales volume that Du Pont and a combined share of the international chemical market that dwarfed that of Du Pont and exceeded that of all the American chemical companies put together. Just as American companies like Du Pont and Dow had moved into direct investment in Europe and Japan in the 1960s, so the Germans in the following decade established subsidiaries in the United States whose total sales equaled almost 50 percent of Du Pont's in chemicals. While Du Pont and the German "big three" were not in direct competition in all fields—the Germans have traditionally been involved in pharmaceuticals, an area Du Pont eschewed, and in turn stayed out of the synthetic fibers field—the probability of confrontation in biochemicals is strong, and Du Pont faced substantial competition in its effort to expand in agricultural chemicals from Bayer and the Swiss firm, Ciba-Geigy.

While Du Pont was girding itself for battle with the Germans in international markets, it faced troubles at home, not all related to the erosion of the synthetic fiber field and its technological somnolence. While Du Pont chiefs saw themselves as struggling to revive from a bad decade and facing growing foreign competition, to many other Americans Du Pont seemed a classic example of a corporate juggernaut, deploying its massive technical, legal, and financial resources to control prices and overrun domestic competitors. In 1979–80, the Federal Trade Commission reviewed in exhaustive detail charges that Du Pont had devoted the preceding six years to a scheme to acquire 65 percent control of the U.S. market in the production of titanium dioxide, a pigment used in white paint, by 1982, and held a 51 percent market share by 1979. To achieve this end, Du Pont had refused to license its lower cost technology and maintained prices at a level "high enough to fund expansion but low enough to discourage competitors from expanding." By so doing the company allegedly planned not only to extract "higher than competitive profits" after its position was assured in 1982, but had blocked the introduction of more energy efficient processes in the inflationary years of 1973–79.[3]

In this particular case, the F.T.C. found the evidence against Du Pont unconvincing, and dismissed the charges. But the episode was symptomatic of the chronic antagonism between Du Pont and critics of big business. Du Pont has the dubious distinction of being one of the companies most frequently investigated and prosecuted for antitrust and related violations since the passage of the Sherman Act in 1890, the target of major divestiture proceedings in 1911–13 and again in 1957–62. In more recent years the antitrust chorus had been joined by spokesmen for the consumer movement such as Ralph Nader, whose researchers produced a lengthy tract on Du Pont's involvement in the political and economic affairs of Delaware (and elsewhere) entitled *The Company State*.[4] By the middle of the decade the company felt so beleaguered that it departed from a century-old tradition, appointing as president Irving Shapiro, a lawyer with no family connections who had helped maneuver the company through a forest of antitrust and related lawsuits in the preceding years. Although some observers in the industry held Shapiro responsible in part for Du Pont's declining commitment to research and innovation during his term as president, he proved to be a vigorous and visible spokesman for Du Pont and the business community in general.

Perhaps even more subversive of Du Pont's view of the world was the emergence of a systematic critique of the environmental effects of the chemical industry. Beginning in 1962 with the writings of Rachel Carson and others on the results of extensive use of pesticides on other life forms, within two decades the environmentalists had stimulated a widespread questioning of the ultimate costs of unrestrained industrial growth, and supported a variety of government regulations on occupational health and safety, pollution abatement, and related matters that Carl Gerstacker, former chairman of Dow Chemical Co., regarded as a reflection of "how much trust the government . . . doesn't have in the industry's capacity to act in the public interest."[5] While Du Pont was contemplating the purchase of Conoco, which produces various chemicals, including controversial polyvinyl chlorides, the newspapers were filled with stories of the havoc wrought at Love Canal on the Niagara River by the chemical wastes left by Hooker Chemical Co. many years before, and congressional investigations of claims brought by former American servicemen suffering from the effects of Agent Orange, a defoliant produced by Dow that was used in the Vietnam War.

Du Pont had maintained a fairly good record in maintaining safety precautions and reducing industrial pollution in recent years. It began programs in "pollution abatement" as early as 1938. But the fundamental arguments of environmentalists extended beyond these specific issues. The intensive use of synthetic soil nutrients, synthetic detergents, and nonbiodegradable chemical byproducts as well as pesticides and herbicides threatened the natural ecological system, while the development of genetic manipulation, whatever its benefits, could pose unforeseen dangers to the survival of the human species. Ultimately, the environmental view of the world challenged the sanguine aspirations of previous generations, a vision embodied in Du Pont's slogan emblazoned on the entry to its headquarters in Wilmington: "Better things for better living through Chemistry"—a slogan the company, perhaps significantly, decided to phase out in 1972.

The concerns that preoccupied Du Pont's executives in 1980–81 have been factors that influenced the company since 1800 when Pierre S. du Pont and his family established a powder mill on the Brandywine Creek outside Wilmington, Delaware. Over the next one hundred and eighty years the company became the major American producer of a range of explosives used in war and peace, then diversified into chemicals after World War I, and by the middle of the twentieth century had evolved into one of the world's leading manufacturers of synthetic fibers and plastics as well as thousands of other products. Concurrently, Du Pont was transformed from a family firm employing a few dozen workmen into a multidivisional corporation with over 230 million common shares outstanding, of which Du Pont family members held about 22 percent, employing more than 130,000 people in hundreds of plants, research labs, and distributing outlets throughout the world.

The Du Pont saga may be seen as a reflection of, and contributor to America's dramatic economic growth over the past two centuries. Du Pont was also a major element in the development of the international chemical industry, and this book proposes to reexamine the history of the company in that context as well. Consideration of the technological, commercial, and organizational transformations of the chemical industry is essential for an understanding of the significance of Du Pont in the American industrial economy.

CHAPTER ONE

The Transformation of the Chemical Industry, 1780–1910

The boundaries of the chemical industry defy precise definition. Spokesmen for the industry like to demonstrate how the products of their operations permeate our daily lives: "Today a stylish woman can go to the theater completely attired in synthetic material, from her acrylic wig to her vinyl shoes. Tomorrow a businessman may be able to leave his all-plastic house in the morning, walk across a lawn of polyethylene grass to his fiber glass car, and drive to work over roads cushioned with synthetic rubber."[1] The ubiquitousness of chemical products and processes, however, creates problems for anyone seeking a broad view of the industry as a whole.

Even chemical manufacturers do not have an entirely clear idea of the scope and limits of their markets, which in any case are constantly being reshaped as new products are introduced. Synthetics replace natural materials, new combinations of synthetics replace existing ones, and new processes spawn unexpected byproducts. In 1930 representatives of Du Pont approached their counterparts at Standard Oil of New Jersey to express concern over the intended operations of a subsidiary of the latter company, Hydro Products, in developing synthetic fuels through hydrogenation of coal. Du Pont had nothing to fear, Standard Oil's Frank Howard reassured them. His company had no desire to "go into the chemical industry." In the fashion of the day the two firms negotiated a mutually satisfactory dividing line between their respective fields. Yet the realm of technology into which Standard Oil was advancing embraced far more than the "fuels and lubricants" defined in its agreement with Du Pont.

Over the next few years, in addition to synthetic rubber, Standard Oil's researchers would develop a synthetic toluene as a substitute for the coal-based product used in Du Pont's staple explosive, T.N.T.; a material called methyl ethyl ketone used in coating fabrics, another established Du Pont market; and an assortment of related chemical products including insecticides, incendiary bombs, and industrial alcohols.[2] After World War II the growth of the petrochemical field extended the overlapping of the two industries and encouraged even small oil companies like Ashland Oil to diversify into chemicals.

Today the dividing line between petroleum and chemicals is difficult to determine and the same can be said for the relationship between chemicals and textiles, chemicals and pharmaceuticals, chemicals and explosives, and chemicals and light metals. Practically all systems of classification are arbitrary. The United Nations Standard International Trade Classification includes explosives with chemicals but excludes synthetic fibers, a field which historically has been tied directly to the chemical industry.

Despite the amorphous and changing character of the chemical industry, certain specific fields have predominated in different phases of its development, in terms of market size and levels of investment, and equally important, in terms of technological and organizational innovation. In the early nineteenth century soda ash and alkalis—the "heavy chemicals"—were most important, joined later by dyestuffs: all three products were tied directly or indirectly to the workhorse of the Industrial Revolution, the textile industry. In the early twentieth century synthetic ammonia, used in fertilizers and explosives, rapidly developed as a major field. After World War I synthetic fibers began to develop on a commercial scale, stimulating research and development in polymers and expansion of plastics and resins that were a central element of the industry during and after World War II.

Traditionally most chemical companies have specialized in a few related products, focusing either on costly and technically complex materials such as dyes or drugs and catering to a small but durable market; or producing large quantities of easily processed materials, such as acids or ammonia, for larger but fluctuating markets. Given the remarkable range of market structures in chemicals, no one company has been able to encompass all of the thousands of products and byproducts of the industry. Some firms, however, have developed the investment capabilities, scale of operations, organizational flexibility, and high technical standards that enabled them to diversify over a broad range of products and establish a significant market position in the industry as a whole. These companies—Du Pont in the United States, Imperial Chemical Industries in Britain, I. G. Farben and its successors in Germany—are the focus of this book.

The Traditional Industries: Alkalis and Acids

Chemical processes were used in manufacturing long before anyone understood the basic principles underlying them. Ores were smelted for metal, limestone was burned to furnish mortar, clay pots were baked and glazed, and various minerals were ground into pigments for paint and heated to produce acids for solvents in the ancient civilizations of the Near East and the Mediterranean basin. Metallurgical processes were refined and extended in Europe during the Middle Ages, and in the era of exploration and commercial expansion that followed, new natural sources of dyes were introduced to supplement the handful of colors mentioned by Pliny in the first century A.D. Trade in chemical products such as dyes, acids, and alkalis increased through the seventeenth and eighteenth centuries as new uses for them were discovered. But the supplies of these products and the consumer goods to which they contributed, such as soap, glass, medicines, and metal fabrications, were necessarily restricted by the inability of craftsmen using traditional empirical procedures to improve production methods or develop substitutes for materials in short or diminishing supply.[3]

Two developments in the latter part of the eighteenth century laid the groundwork for the modern chemical industry: the application of systematic inquiry and experimentation to explain chemical reactions, and the rapid growth of demand for chemical products by the textile and iron industries among others, "a demand which could be satisfied only by production in tons of what was formerly made in pounds."[4] The steam engine in the nineteenth century enabled chemical manufacturers to expand production to meet these demands while incidentally reducing the costs of extraction and transportation of raw materials.

The chemical industry of this era was by no means a "big business" by our standards. As late as 1852 British alkali makers, although they employed 6,000 workers in total, averaged only 200 per firm, with a total capital investment of $3.5 million, approximately one fiftieth the size of the textile industry in that year. But chemical manufacturers occupied a strategic position as suppliers of necessary intermediates to the larger consumer-goods producers, and the opportunities for profits were substantial, at least up to the point where the technology of production became widely known and new entrants saturated the market, a process that troubled industrialists throughout the era.[5]

France and Britain share the honors in the beginnings of modern chemical research but Britain shortly became the center and beneficiary of the application of theory and experiment to commercial development. In the 1770s the French chemist and public servant Antoine-Laurent Lavoisier initiated experiments, later extended by Joseph Priestley and Henry Cavendish in England whose work culminated in the identification of hydrogren and oxygen as basic chemical elements, and provided the conceptual framework for the systematic discovery and classification of other elements over the next one hundred years. Lavoisier also pioneered in the field of state-supported technical education: one of his students in the art of powder-making was Irenee du Pont who worked at Lavoisier's laboratory at Essone in 1785–88.[6]

Lavoisier and other chemists associated with the old regime in France were guillotined during the Revolution in 1794, but the revolutionary government subsequently resurrected the tradition of state support for the sciences with the foundation of the Ecole Polytechnique and subsidies for research. Encircled by enemies, the French government under Napoleon encouraged the rapid development of industries contributing to military needs and domestic food supplies. Cost effective methods of production, however, were sacrificed in the intense effort to ensure national self-sufficiency, rendering the emerging French chemical industry vulnerable to competition and reliant upon government duties and import embargoes to protect it from more efficient firms emerging on the other side of the English channel.[7]

Even before the Revolution the French government was concerned about the dependence of many industries, particularly soap, glass, paper, and porcelain, on supplies of alkalis derived from potash produced from leached wood ashes that had to be imported from the forests of Russia and Scandinavia, or from barilla, ashes produced from seaweed imported from Spain. The Academie des Sciences in 1775 offered a reward to the inventor of a process that would rely on salt, a raw material domestically more readily available. During the 1780s Nicholas Leblanc, a private surgeon to the Duc d'Orleans, developed a method of extracting sodium carbonate or "soda ash" as a substitute for potash, combining salt with sulfuric acid and heating the solution with limestone and coal. Leblanc never received the prize, and the patent he was awarded in 1791 for his process, along with the factory he established with capital advanced by the Duc d'Orleans, was confiscated by the revolutionary government in 1794. The factory was returned to Leblanc in 1804 but he was unable to raise new capital and committed suicide four years later. Ironically, at this point Napoleon, cut off by the British from Spanish barilla, encouraged the production of soda ash based on the Leblanc process. Leblanc's Paris factory was acquired by the emperor's pharmacist and other soda producers sprang up near the center of the French soap industry at Marseilles. Between 1809 and 1816 production of soda ash in France rose from 2,000 tons to 13,000 tons per year, doubling again in the following decade.[8]

In England the search for new sources of alkalis paralleled developments in France. Although the French, using the Leblanc process, were the first to produce soda on a commercial scale, the British were more advanced in certain related fields. In 1746 John Roebuck of Birmingham, whose later coal-mining operations led to the development of the steam engine by his business associate James Watt, developed a method of burning sulfur and oxygen in corrosion-resistant lead chambers to produce sulfuric acid in much larger quantities than had been possible before. In 1749 Roebuck established at Prestonpans in Scotland "the first substantial chemical enterprise in the world," but failed to patent his lead chamber process for more than twenty years, by which time it had spread to France. In Roebuck's time sulfuric acid was primarily used in bleaching cloth and cleaning metals, but it was also an essential ingredient in the Leblanc soda process, and the two industries were to be closely linked in the nineteenth century.[9]

The main impediment to the growth of the artificial soda industry in Britain through the period up to 1822 was the duty on imported salt that deterred most would-be soda makers even though shortages of natural alkalis kept prices high on soaps and bleaches. The conservative attitudes of alkali consumers toward the artificial product, a recurrent problem for British chemical innovators, also affected the situation. In 1820 British soda and sulfuric acid output was only about one sixth the size of the French industry.[10] During the next three decades, however, the position of the two countries was to be reversed. By 1860 British alkali manufacturers were producing 160,000 tons per year, four times the levels of their French competitors, and British soda and bleaching powder exports dominated the continental European and American markets.

There were several reasons for this dramatic transformation. First, industry in general developed more rapidly in Britain than elsewhere in this period. Strategically positioned as suppliers to the glass, soap, leather, and paper industries, alkali manufacturers also benefitted from the expansion of coal mining and transportation improvements that created a larger domestic market. Furthermore, chemical manufacturers sought new uses for soda ash and its byproducts, expanding markets while reducing production wastes, albeit sometimes in grudging response to public and government pressure rather than due to commercial foresight.

Another characteristic of the British soda manufacturers was their recognition of the commercial possibilities of technical and organizational improvements. One of the largest soda producers, Charles Tennant, anticipated his competitors by integrating his business with sulfuric acid production as early as 1803, thus reducing production costs. Tennant also took the unusual step of integrating into marketing rather than relying on independent distributors, but here he had few imitators. Up to the 1880s most British soda makers operated on a small scale and avoided direct marketing.[11]

The circumstance that stampeded Leblanc soda producers to pool their resources and move toward vertical integration, as well as develop further uses for their byproducts, was the challenge presented by a new process for soda production pioneered by two Belgian brothers, Ernest and Albert Solvay, in the 1860s. The Solvay method involved the combination of salt with ammonia and carbon dioxide. The "ammonia soda" process had several advantages over the Leblanc process: soda could be produced much more rapidly and inexpensively as fewer steps and ingredients were involved; a greater amount of useable soda could be obtained from the salt; and virtually all the ammonia could be recovered and recycled. One estimate places the cost of one ton of soda by the Solvay process at two thirds the cost of one ton produced by the Leblanc process.

Unlike Roebuck, who tried to patent his sulfuric acid process too late, and Leblanc, who lost the patent to his development, the Solvays carefully covered every step of their way with patents in other countries as well as Belgium, a practice the German dye makers and the Du Ponts were to emulate. Having established their position technologically, the Solvays then proceeded to secure it by licensing their patents to selected foreign manufacturers, assisting them financially as well in exchange for a strong interest in the stock of the new companies. This assistance was particularly important as the Solvay patents did not include crucial information on manufacturing procedures and the Solvays preferred to deal with small entrepreneurs who needed their capital resources and know-how rather than with established firms. The Solvays used these techniques to exercise indirect control over the ammonia soda field as a whole, ensuring that they would always have access to new technical developments and, more significantly, preventing or at least restricting the proliferation of small competing firms that characterized the Leblanc soda field and many other industries of the time. Their emphasis on stability and informal cooperation among soda producers in the international market left a strong imprint on the chemical industry as a whole, as did the parallel efforts by Alfred Nobel and his successors in explosives.[12]

In 1872 the Solvays licensed their process in Britain to John Brunner and Ludwig Mond, an emigrant who had studied chemistry in his native Germany. The Brunner-Mond enterprise was exactly the kind of small but technically proficient venture the Solvays preferred to promote, and over the following decade Solvay capital carried the company through a difficult gestation period as increased soda production by Leblanc manufacturers drove alkali prices down. By the early 1880s the Brunner-Mond firm began to show profits and entered the export market, focusing on the United States. The Solvays, who controlled 20 percent of Brunner-Mond at the time, in 1885 agreed to give the British firm an exclusive market in North America as well as Britain, although this arrangement was complicated by the fact that Solvay had already licensed another soda producer in the United States.[13]

By this time the British Leblanc soda makers were steadily losing ground to the Solvay companies, not only from Brunner-Mond in their domestic markets but also to other Solvay affiliates on the Continent. In 1890 the British Leblanc companies merged together around Tennants, the largest of their group, to form the United Alkali Company Ltd., with a total soda capacity of 140,000 tons per year and a capital stock issue of $40 million, making it "the largest chemical enterprise in the world." But the new firm was unwieldy, its management obliged to mediate among the various formerly family-controlled companies while seeking to preserve an outmoded technology. Within three years United Alkali had been outstripped by Brunner-Mond and had to depend on its sales of byproduct bleaching powder and caustic soda to avoid disaster.[14]

In Germany low tariffs on alkalis had discouraged the growth of a domestic soda industry up to the 1860s and there were few Leblanc producers to resist the ammonia soda process, incorporated by the Deutsche Solvaywerke in 1885, which within two years produced one third of that country's alkalis. In France, on the other hand, two large Leblanc soda makers had emerged by mid-century: Kuhlmann and St. Gobain. Kuhlmann was involved in a variety of chemical fields including phosphates for fertilizers and St. Gobain had entered the soda field primarily to supply its needs in glass production. Neither company was particularly large by British standards and exports were small. French chemical entrepreneurs were also reputed to be financially conservative and technologically somnolent, factors that had contributed to the decline of the French chemical industry after 1830. But other factors may have been more important than the internal failings of French manufacturers, in particular the slower rate of French industrial growth generally compared to Britain, Germany, and the United States. Du Pont, whose origins are French, also acquired a reputation for financial caution, technological imitativeness, and a reluctance to enter export markets. In any case it is perhaps not surprising that later the Du Ponts preferred dealing with like-minded French businessmen and increasingly staid British chemical manufacturers to the aggressive and venturesome Germans.[15]

Germany and the Dyestuffs Revolution

Although German chemical manufacturers produced as diverse a range of materials as any other European country in the late nineteenth century, in the public mind the German chemical industry was virtually synonymous with dye-making and the big German chemical firms acquired the not undeserved sobriquet the "dyestuff monopoly." In 1913 Germany produced 295,000,000 pounds of dyes and was estimated to control 88 percent of the total world market in the field. The German dye business was also "big business" at least in comparison with other chemical enterprises of the time. In 1904 the Ludwigshafen factory of the Badische Anilin und Soda Fabrik (B.A.S.F.) company alone employed over 8,000 people, covered 540 acres, and was valued at $19 million. Two years earlier the entire assets of the Du Pont Powder company, the largest explosives producer in America, were valued at between $12 million and $14 million and employed fewer than 5,000 workers. B.A.S.F. was only one of the six large German dye firms that moved steadily from the 1890s on toward greater consolidation, culminating with the establishment of the I. G. Farbenindustrie in 1925.[16]

The dyestuffs industry grew as rapidly in Germany in the last half of the nineteenth century as alkalis had in Britain in the preceding fifty years, but with more enduring results. Even today the German companies produce more dyes than any other country, including the United States, and their earnings in 1978 exceeded those of American dye makers by $350 million. Furthermore, research and development in dyestuffs propelled the Germans into related fields including pharmaceuticals, plastics, synthetic nitrogen, and synthetic fuels, and they have maintained the technical and commercial edge in most of the fields they entered.

The dyestuffs industry was a direct outgrowth of research in organic chemicals in the early nineteenth century, an area in which German scientists, in particular Justus von Liebig, predominated. Liebig also established one of the first modern chemical research laboratories at the University of Geissen in the 1820s, emphasizing systematic and exhaustive experimentation. Liebig's methods were transmitted to England two decades later through one of his students, August Wilhelm Hofmann, who took a position, at Liebig's urging, at the Royal College of Chemistry in London in 1845.

During this period German chemists were investigating the peculiar characteristics of chemical compounds derived from carbon, the basic element in all organic matter. The carbon compounds were exceedingly complex and soon became a separate field of research, designated organic chemistry. Before moving to London, Hofmann had begun research in this field, investigating the properties of coal tars. In 1856 he set one of his students in London, William Perkin, the task of synthesizing quinine from napthalene deposits. This coal tar residue left by gas works was in plentiful supply in Britain while cinchona, the natural source of quinine, was rare and had to be imported. Perkin's experiments with napthalene were unsuccessful—quinine was not to be synthesized until 1944—but while working with another carbon compound, aniline, Perkin discovered a substance that produced a rich purple dye, later called mauve.

Perkin's father provided capital to establish a dye business for his son in 1857. The younger Perkin was able to assemble his raw materials cheaply from the coal tar deposits of gas works, treating them with hydrochloric acid to produce aniline. Despite the usual initial resistance of textile makers to the new synthetic dye, the demand for Perkin's product increased rapidly after he developed processes to apply and hold dyes on cotton and wool as well as silk cloth. Before the end of the 1860s, sometimes called the "Mauve Decade" in honor of Perkin's discovery, a larger variety of dyes had been identified, including the family of "azo dyes" produced by nitrous acid reacting on aniline, which could be applied directly onto cloth.[17]

Despite their early lead in the field British dye manufacturers soon lost momentum. Research in organic chemistry languished after Hofmann returned to Germany in 1865. Perkin himself retired from the dyestuffs scene in 1874 after developing synthetic red alizarin, an extremely versatile dye that opened the way for discovery of a range of other synthetic substitutes for natural dyes, most notably indigo, developed by Adolf Baeyer for B.A.S.F. in 1897 after more than thirty years of research.

Perkin reemerged on occasion to publicly criticize his former colleagues in dye-making for failing to promote trained chemists to managerial positions in their firms and neglecting to support chemical education generally. Another problem for the British dye manufacturers was that banks were reluctant to finance their ventures which operated in volatile markets influenced by the whims of fashion and required longer-term financing than banks customarily provided, in order to develop new processes and find markets for them. Finally, the inexpensive sources of coal tar that had enabled Perkin to produce dyes commercially diminished as gas works in the 1870s began to develop more efficient methods of recovering their wastes, and found it more lucrative to export distilled tar to the growing German market.

By the 1880s the British dye industry was in serious decline. An exception to this trend, however, was a company established in 1864 by Ivan Levinstein who, like Ludwig Mond, was an emigrant from Germany, where he had studied organic chemistry. Levinstein manufactured his own intermediates and fostered research. When the Du Ponts decided to enter the dyestuffs field in World War I they first approached Levinstein for technical assistance.[18]

In France as well a strong initial position in organic chemicals was lost, largely as a result of "l'affaire Fuchsine" in 1863, in which a company established to exploit the fuchsin dye, which was also an important intermediate in producing other dyes, was awarded a monopoly under the French patent laws despite the fact that its product was inferior to others using more efficient processes. Other would-be French dye makers departed, principally to Switzerland, and La Fuchsine shortly went bankrupt as it was unable to compete with imported, or smuggled, products, but retained the monopoly up to 1870.[19]

Germany's success in the field was thus partly the result of the weakness of potential competitors in other countries. But there were also important positive factors that contributed to and sustained the predominance of the German organic chemical industry. The most important element was their emphasis on research, integrated into commercial operations not only in the development of new dyes but also in improving production techniques, assessing the relative value of new materials, and designing plants. In the alkali and sulfuric acid fields this kind of constant monitoring had been unnecessary since the industrial processes were straightforward, but in dyestuffs procedures were more sophisticated and required trained chemists. Furthermore, as competition increased in existing lines, the development of new dyes became the major factor in market leadership, foreshadowing trends in all the advanced technology industries in the twentieth century. Finally, German industrial chemists were able to develop new methods of extracting tars from lignite or brown coal which was abundant in Germany, so that by the 1890s they were no longer dependent on imported tar distillates.

The German dye industry pioneered in systematic industrial research. In 1868, only three years after its establishment, B.A.S.F. recruited Heinrich Caro, who had studied in London under Hofmann, to set up a research laboratory at Mannheim. B.A.S.F.'s lead was followed shortly by Farbwerke Hoechst of Frankfort, one of whose founders, Eugen Lucius, was himself a trained chemist. By the 1880s all of the major German chemical companies were supporting research on a large scale. Moreover, unlike the British, the German companies encouraged the movement of research chemists into senior management. The most notable early example of this practice was the career of Heinrich von Brunck, one of Caro's early associates at B.A.S.F., who later became technical director and chairman of the board of that company. Brunck in turn promoted another young chemist, Carl Bosch, who became chairman of B.A.S.F. and chief executive of I. G. Farben in the 1920s. Equally spectacular was the career of Carl Duisberg who developed a range of dyes for the Friedrich Bayer concern in the 1880s, then overhauled Bayer's whole research operation to make it competitive with Hoechst and B.A.S.F., and became chief executive in 1911, by which time Bayer was the largest chemical company in Germany.

Brunck, Bosch, and Duisberg were only the most remarkable members of a large group of technically trained businessmen in the German chemical industry. By 1900 Germans were patenting new developments at a rate four times greater than that of the British in chemicals. The German dye firms had also diversified, not only into fields related to coal-tar derivatives such as pharmaceuticals, but also into the traditional soda and sulfuric acid markets where they were introducing new processes to augment the recovery and recycling of byproducts, and into novel fields such as celluloid film and plastics.[20]

Technical leadership was a crucial factor in the success of German dyestuffs, but it was not the only one. The unified German government after 1870 adopted policies to protect domestic manufacturers and promote exports, including dyestuffs and related chemical products. The most important contribution of the government to the German dye industry was the patent law of 1877 which provided for a thorough search before issuance of a new patent, reducing time-consuming and expensive litigation, and ensuring that a patent once issued was virtually impregnable. In 1891 the law was revised to allow the patenting of "refinements" to a process, an arrangement that particularly suited the chemical industry where research was constantly improving existing processes.

The German dye makers also followed the lead of the Solvays in alkalis, taking advantage of the establishment of patent laws in all the industrial nations that protected foreign as well as domestic patentees. Britain and the United States in particular offered fruitful fields, for their patent laws did not require that a new patent be put into production or licensed. As a result German and other foreign manufacturers could patent a wide range of processes and then hold these markets captive to their exports for the life of the patent. Unlike the Solvays the German chemical firms avoided licensing or foreign direct investment as much as possible in this period, partly from fear of losing their technical edge over foreigners who might pirate their processes.

In addition to their manipulation of foreign patent laws, German chemical makers acquired a reputation for sharp commercial practices. Would-be competitors in the United States complained that the Germans cut prices in specific countries whenever faced with new rivals abroad. Textile manufacturers confirmed that the Germans employed techniques such as "full line forcing" in which a buyer had to accept an entire line of dyes from the German exporter even if cheaper dyes were available elsewhere. Instances of bribery were cited as evidence of unethical business practices, although hardly unique to the Germans at this time.

After 1885 the German dye exporters began to integrate into marketing, replacing wholesale agents with small sales companies abroad manned by trained technical consultants, and linking research in new products to trends in consumer demand in textiles and related fields. While German dye salesmen did use the techniques highlighted by their critics, the main source of their success in foreign markets was sophisticated marketing combined with technological efficiency and innovation.[21]

The German chemical makers also developed much stronger ties with banks than their counterparts in other countries. In contrast to the wariness of British and French bankers even toward established chemical companies in such fields as alkalis and acids, the Frankfort and Rhenish banks provided 45 percent of the capital for expansion of the German dye industry between 1880 and 1900. On the whole the investment proved worthwhile: average dividends on shares in the four leading German dye companies ranged from 20 to 24 percent from 1890 to 1897, a period marked by general economic problems, and in the following decade the average rose to 30 percent. The increasing connection with the banks related to another unusual characteristic of the German chemical industry in this era, the rapid transformation of the companies from private to public enterprises with widespread distribution of stock.[22]

This development probably facilitated the next step, the progressive consolidation of the German organic chemical industry. In the 1860s there were more than thirty companies in dyestuffs in Germany, but twenty years later only six firms of consequence occupied the field. The largest were B.A.S.F. and Bayer, followed by Hoechst and Agfa (Aktiensgesellschaft fur Anilinfabriken of Berlin), the last a company formed in 1867 by the prominent chemist Carl Martius and Paul Mendelssohn-Bartholdy, banker son of the famous composer. On a much smaller scale but also involved in dye exports were Kalle and Company of Bieberich, and Cassella and Company of Frankfort.

Despite their growing strength in export markets, the German dye makers all faced the dilemma of maintaining heavy capital investment for continuing research while earnings were declining due to overproduction. This was not an unusual situation: many other industries confronted similar problems in the United States as well as Europe in the late nineteenth century. In fact the initial solution to the dilemma, cartelization through industrial trade associations, seems to have been borrowed directly from the American industrialists, such as Du Pont in the explosives field, who were groping for methods of market stabilization from the 1860s on. Between 1878 and 1904 German dye concerns negotiated a variety of agreements among themselves to stabilize prices and allocate markets in the growing range of dyes. Eventually these agreements included patent exchanges and mutual pledges to refrain from entering new fields in competition.

Like the various pools and conventions established by American manufacturers in this period the German dye cartels were usually short-lived and unstable; but where in the United States the Sherman Act of 1890 established legal barriers to cartelization, in Germany cartels acquired a stronger legal foothold and eventually were to be encouraged by the government as a means of consolidating resources in order to move effectively into export markets. These differences in tradition were to have a marked effect on the structure of the chemical industry in the two countries in the twentieth century.

By 1900 the lengthy development of synthetic indigo had strained the resources of the largest German dye firms and the market for new dyestuffs seemed to have peaked. As the German businessmen cast about for solutions to their difficulties, American innovation in business organization again had a significant impact. Following an American trip in 1903, during which he had observed the development of new American corporate giants such as Standard Oil and United States Steel, Carl Duisberg of Bayer proposed a merger of the six big German dye companies which would enable them to reduce costs of research and manufacturing, eliminate conflict over patents, and reduce the rate of distribution of earnings to make more funds available for reinvestment and diversification. Duisberg's plans for consolidation were stymied, however, by Hoechst which had expansion plans of its own. Hoechst entered an arrangement with Cassella that involved an exchange of shares between the two companies and an interlocking directorate, a business structure known as an Interessengemeinschaft (I. G.). In 1904 Bayer joined with B.A.S.F. and Agfa in a "Dreierbund" involving a pooling of patents and redistribution of earnings, to counter the Hoechst-Cassella alliance. B.A.S.F. and Bayer were the senior partners, receiving 43 percent each of the profits while Agfa received 14 percent.

Neither arrangement constituted a merger along the lines of the American companies Duisberg had advanced as a model, and the German companies were to continue to debate the relative merits of cartelization versus consolidation. In practice, however, both groups introduced a much greater degree of coordination and integration into their respective organizations than had been anticipated initially, particularly in research and sales. Furthermore, the two alliances entered into price and market agreements with each other in key dyes so that the problems of competition that had troubled Duisberg did not seriously affect the earnings of either group. As far as foreign customers and competitors were concerned, the "German dye cartel" functioned as a single unit, and its progressive amalgamation in 1916 and 1925 was significant primarily in terms of internal management organization.[23]

By the early twentieth century, then, the German dyestuffs makers were moving toward a coordination of operations that would give them as strong a position in this field as the Solvays held in the alkali industry and the Nobel group in explosives. For all of these European chemical manufacturers the largest single market lay across the Atlantic in the United States. The American market, already absorbing over $800 million in chemicals, was a booming, growing economy with an expanding population whose rising living standards portended a tremendous future. Whoever ultimately developed this market could well dominate the international chemical industry.

The Chemical Industry in America

The domestic American chemical industry shared in the rapid growth of the national economy after the Civil War. Total capital investment in chemicals increased from $206 million in 1880 to $1.5 billion in 1910 in constant dollars, and the total volume of chemicals produced increased more than ten times over in this period. The industry was relatively small compared to steel or textiles, each of which had assets of more than $5 billion on the eve of World War I, but it represented dramatic growth, achieved almost entirely through re-investment of earnings as bankers here as elsewhere, except Germany, "shied away from chemicals, aghast at the . . . quick obsolescence of chemical processes and apparatus."[24]

The American chemical industry exhibited certain peculiar characteristics that reflected the circumstances of its growth but also the structure of the international chemical markets. Outside of Du Pont in explosives and the General Chemical Company in sulfuric acids, the largest American chemical firms were those involved in the relatively simple processes of manufacturing fertilizers from phosphates. Three firms dominated this field by 1900, producing a combined output of almost 3 million tons, about 25 percent of world production, to serve the growing demands of American cotton farmers. They did not attempt vertical integration, relying instead on outside suppliers of sulfuric acids and competing with numerous smaller companies in local markets. They were overcapitalized and ran into serious financial difficulties when fertilizer demand fell off after World War I and new sources of nitrogen-based fertilizer expanded in the 1920s.[25]

Neither in dyestuffs nor alkalis, the main fields of European chemical endeavors, were American producers of any real consequence before World War I. This situation was largely because the Europeans established an early foothold in American markets. In 1869 Jacob Schoellkopf of Buffalo, New York, established a successful dyeworks that soon became an integrated company, producing its own intermediaries. Schoellkopf's success attracted other entrepreneurs, but American textile manufacturers in 1883 pushed through tariff revisions that cut duties on imported dyes to 35 percent *ad valorem,* opening the way for German and Swiss dye exporters who consolidated their market position by taking out a wide range of American patents. In 1914 the United States imported more than 90 percent of its dyestuffs and, except for Schoellkopf, most of the factories were "little more than assembly plants . . . entirely dependent on intermediates imported from Germany."[26]

A fairly substantial sulfuric acid industry developed by the early nineteenth century. One of the first acid producers in the country was Harrison Brothers of Philadelphia which later went into paints and pigments and for that reason was acquired by Du Pont in 1971. In New England, textile manufacturers produced their own sulfuric acid for bleaching while in the Middle West, Eugene Grasselli, an Italian chemist trained in Germany, established an acid and diversified chemical enterprise in the 1840s, expanding rapidly to serve the regional oil refining and fertilizer markets. Grasselli was typical of American chemical manufacturers in pursuing expansion through product diversification rather than specialization or vertical integration. An exception to this generalization was William H. Nichols, also a sulfuric acid producer, who in 1899 arranged the merger of twelve firms into the General Chemical Company with assets of $14 million, roughly equal to Du Pont's at this time. Nichols, a trained chemist, also sought to emulate the Germans by establishing a research laboratory and buying foreign patents or licenses.[27]

Despite the abundance of sulfuric acid, the Leblanc alkali process did not become important in the United States, primarily because of inexpensive imports from Britain. As in Germany the alkali business in America appeared with the introduction of the new ammonia soda technology. The Belgian Solvays were among the first to exploit the opportunities of the American market. In 1881 Rowland Hazard of Providence, Rhode Island, negotiated an American license arrangement with the Solvays similar to that acquired by Brunner-Mond in Britain ten years earlier. In 1884 the Solvay Process Company was set up in Syracuse, New York, with capital and technical help from the Belgians who held 50 percent of the stock. Within three years the Solvay Process Company had acquired almost one quarter of the American alkali market, at the expense of the British Leblanc producers and Brunner-Mond.

Brunner-Mond was not pleased with the appearance of a rival in one of its most lucrative markets and, after some bickering among the Solvays and their licensees, the British company agreed to reduce its exports to the United States in exchange for shares in the Solvay Process Co. Between 1889 and 1905 Brunner-Mond steadily pulled out of the American market and Solvay Process expanded its production, buttressed by increased tariffs on imported soda ash in 1890 and 1897, measures which incidentally virtually ensured the rapid decline of the British Leblanc combine, United Alkali Company.

Neither Brunner-Mond nor Solvay Process, however, recognized the significance of the new electrolytic processes for the production of alkalis that became commercially feasible around the end of the nineteenth century. Various small companies in the United States, relying on the Castner electrolytic patent, began to enter the soda and bleaching field so that the market position of Solvay Process began to deteriorate despite increased sales. When the British and European shareholders in Solvay Process sought to intervene in its management to improve its competitive position, the American minority group resisted, and internal rivalry in the Syracuse company weakened it further. Solvay Process and its coke supply affiliate, Semet-Solvay Company, were eventually to be swallowed up in the giant Allied Chemical and Dye merger at the end of World War I. The Brunner-Mond investment in the American company would later significantly influence negotiations among international chemical companies in the 1920s.[28]

In summary, American chemical producers in th early twentieth century, outside of the phosphate fertilizer field, operated on a relatively small scale in family-owned firms serving local markets and financing such growth as they experienced largely from internal sources. Only a few, such as Schoellkopf's dye enterprise and Nichols's General Chemical had attempted vertical integration to any great extent. At this point, however, the Du Pont powder company began to develop the technical, financial, and organizational innovations that would enable it to move to the forefront of the American chemical industry during World War I and would provide it with the resources necessary to bargain with the large foreign enterprises on equal terms.

CHAPTER TWO

Du Pont: The Rise of the Explosives Company, 1801–1913

The explosives industry was transformed by changes in technology and organization in the nineteenth century to a greater extent than any other chemical field except perhaps dyestuffs. The development of dynamite, smokeless powder, the safety fuse, and the blasting cap all significantly affected the military and industrial uses of explosives, thus producing serious dislocations in manufacturing and commercial markets. In 1850 the production of black powder, the main explosive, was fairly straightforward, involving the combination of refined saltpeter, charcoal, and sulfur, all materials that were reasonably accessible to European and North American markets. Some governments maintained powder plants for military needs, but for the most part private producers held the field, operating on a small scale and serving localities or regions. Within the next half century the industry became far more concentrated as changes in chemical technology dictated greater reliance on patented processes, imports of scarce materials, and consequently larger amounts of capital and stable, dependable markets.

In Europe a group of loosely associated companies established by the Nobel brothers of Sweden dominated the explosives markets and allocated among themselves much of the rest of the world. In the United States the largest firm was E. I. du Pont de Nemours Company which exercised control over the American market through its dominant position in the Gunpowder Trade Association. In 1897 Du Pont and the Nobel group negotiated an agreement that in effect encompassed the entire international explosives industry. As in alkalis and dyestuffs, by 1900 the explosives field had acquired a coherent and coordinated structure. Developments in the next two decades would dramatically alter that structure but not the basic characteristics of the firms involved nor the attitudes of the businessmen who controlled these large companies.

Du Pont's premier position in American explosives was less the result of technical leadership than the careful deployment of earnings into strategic investments, systematic development of markets, and cultivation of political and government contacts when necessary. At the turn of the century the Du Pont enterprise, despite its size and apparent solidity, was in a state of internal division and confusion. A triumvirate of cousins took over the firm in 1902, however, rather than see ownership pass out of the family. They were to project it into a phase of unprecedented growth. Their instruments of change included the introduction of new management practices and a strategy of diversification from explosives into general chemicals. At the same time the new leaders displayed many of the characteristics of their nineteenth-century predecessors: caution, shrewdness in assessing new opportunities, and thoroughness in exploiting them. The growth of Du Pont from a family firm in a single field to a diversified corporate organization was not unique in American business in this era, but the degree of continuity and the persistent sense of family and company cohesion were unusual characteristics that provided Du Pont with an advantage in its dealings with other firms at home and abroad. This continuity in attitude and policy was to be a prominent feature of the Du Pont company up to the 1950s.

The Powder Company, 1801–1902

The Du Pont family entered the explosives industry more or less by chance. The "founding father" Pierre S. du Pont de Nemours had been a prominent economist, politician, and civil servant in France, inspector general of commerce under Louis XVI, an associate of the reform minister Turgot, and active in the early stages of the French Revolution. Imprisoned during the Terror and critical of the government of the Directory, Pierre du Pont decided to take refuge in America, planning to invest in land and establish a colony that would promote a balance of agricultural and industrial undertakings in keeping with his Physiocratic beliefs. Various reform-minded French political figures including Lafayette, Louis Necker, and Jacques Biderman promised to support the venture.

This overly ambitious scheme collapsed soon after the Du Ponts emigrated to America in 1800. Casting about for alternatives, Pierre's son, Eleuthere Irenee, discovered that the gunpowder produced in the United States was high priced and of extremely poor quality. As an apprentice in powder making under Lavoisier, Irenee felt confident that a successful venture could be initiated, particularly since gunpowder had tariff protection in the United States. He persuaded the family, some of their French backers, and a few Americans, including the French-born Pierre Bauduy of Wilmington, Delaware, to support the enterprise. A joint stock company with $36,000 capital was incorporated in Paris and New York in 1801. At Bauduy's recommendation Irenee set up a powder mill outside Wilmington on the Brandywine creek, borrowing an additional $30,000 from banks in Philadelphia to build one of the largest explosive plants in the country at that time.

Irenee was the driving force behind the powder venture, benefitting from the various political connections of his family both in France and the United States. From contacts in the French government whose foreign minister, Talleyrand, took an interest in the enterprise, Irenee acquired information about improved methods for refining saltpeter, and this knowledge in turn helped land a contract from the U.S. government to provide saltpeter to the War Department. The company did well in its early years: earnings betwen 1803 and 1810 averaged 18 percent of sales and total assets tripled in the period 1810–15.

Declining postwar markets and a series of disastrous accidents and fires hurt the company in the short run but also enabled the Du Pont family to consolidate its control. For many years Irenee had bickered with Bauduy over management of the company, and in 1815 he took over Bauduy's shares, albeit as a considerable financial burden. The French shareholders were also persuaded to exchange their voting shares for long-term notes. It took Irenee another twenty years, virtually to his deathbed, to pay off these debts, but the company was secured for the family for the next century.

In 1834 Irenee was succeeded by Antoine Bidermann, scion of one of the original French sponsors of the powder venture, who had come to work on the Brandywine in 1814. Bidermann, an accountant, ensured that the remaining debts of the firm were paid by the time he retired in 1837. The company was then taken over by Irenee's three sons and four daughters who operated it as a partnership that was virtually a collective enterprise for the family. Nominally all family members had an equal role in management and share in the profits. Property, including the family homes, was held in common and apportioned according to need. The du Ponts extended this approach to their work force generally through a paternalism that was not particularly unique for the early nineteenth century, but proved to be exceptionally enduring. Irenee had initiated these practices when he established pensions for families of employees killed on the job, a frequent occurrence in the explosives industry, and company housing for workers. Du Pont paternalism had a practical basis as they needed to retain trained workmen in a society which provided many opportunities for skilled and semiskilled artisans, but it also reflected the influence of Pierre du Pont who sought to promote a harmonious society as well as an efficiently productive economy.[1]

The partnership and the practice of collective ownership continued to the end of the century but collective management was less durable. From 1837 to 1850 Irenee's eldest son, Alfred Victor, was de facto chief executive of the powder company. Trained in chemistry by Thomas Cooper, a former associate of Joseph Priestley in England, Alfred exhibited more the temperament of a research scientist than a captain of industry. For twelve years he steered the company through a period of extended depression and growing competition in the American west, subject to constant criticism from other family members. Du Pont provided about one million pounds of powder to the U.S. government during the Mexican War but the sudden expansion of operations set the scene for another terrible accident in the Brandywine mill that unnerved Alfred. He resigned three years later and was succeeded by his younger brother Henry, a West Point graduate who relished the titles "Boss" and "General" and dominated the company from 1850 to his death in 1889.

Henry du Pont and his nephew Lammot were the principal architects of the powder company's growth in the nineteenth century. Lammot, the second son of Alfred Victor, combined business acumen with talent for technical innovation, qualities he would pass on in turn to his sons. In 1857 Lammot patented a process for refining saltpeter from nitrates from Peru which were far more plentiful and accessible to American markets than the existing sources in India. His new product, called "soda powder," proved to be a far more effective source of blasting powder than other grades on the American market, and suited the needs of the growing mining industry. Henry du Pont was quick to exploit the competitive opportunities provided by Lammot's improvements.[2]

Henry also recognized the need to expand geographically to tap emerging markets. In Irenee's time more than one third of the powder produced in the United States was consumed in the Eastern seaboard region or in commercial centers accessible to seaborne trade. by the 1840s canals and railroads were opening up the Middle West and local mills sprang up to serve this area. The same situation developed in California where the gold rush produced a booming demand for blasting powder for mining. In 1859 the Wilmington company established a new mill near Scranton, Pennsylvania, adjacent to the burgeoning anthracite coal and iron industry. During this decade as well Du Pont moved into international markets for the first time, supplying powder to both sides in the Crimean War.

When the Civil War began, the Du Pont mills were responsible for producing one third of the total explosives capacity of the United States. The family,

Rolling mill, lower powder yards, near Wilmington, c. 1865. *Courtesy of the Eleutherian Mills Historical Library.*

whose politics tended toward Republicanism of the Whig variety (Henry du Pont would only smoke a brand of cigar patronized by his idol, Henry Clay) were stalwartly against secession and played a role in holding Delaware, a slave-holding state, in the Union column. Lammot du Pont went to Britain on a quasi-official mission to corner the saltpeter supplies in London on behalf of the U.S. government, most of which was turned over to Du Pont mills for refining.

Du Pont was of course a major supplier of gunpowder to the American government during the war, increasing production three times over between 1861 and 1865, with net earnings of about $1 million. But the wartime boom created serious problems. All of the Eastern powder makers increased capacity during the war but they had lost their position in West Coast markets to local California firms. The U.S. government had excess powder stockpiled at the end of the war and the private companies had to buy it up or face even more constricted demand. Du Pont, however, managed to avoid absorbing more than 10 percent of this surplus. Between 1866 and 1871 explosives confronted the same problems of overcapacity and declining prices that plagued other expanding industries in the late nineteenth century. The powder makers responded in much the same fashion as British alkali manufacturers and German dye makers: in 1872, at the instigation of Henry du Pont, the six major Eastern producers formed a cartel, christened the Gunpowder Trade Association, more popularly called the "Powder Trust."

The three largest Eastern producers, Du Pont, Laflin and Rand, and Hazard Powder Company, controlled the largest number of votes in the association and thus determined the allocation of markets and prices which were set at varying levels for different regions. Within three years the Powder Trust had achieved its initial aim, forcing an agreement on the California companies that divided up the lucrative "neutral territory" of the Rocky Mountain mining states. The organization concentrated thereafter on developing new instruments of enforcement through quotas supervised by a permanent "advisory committee" with Henry du Pont at its head. Independent black powder makers who resisted the advances of the trade association were subjected to fierce price wars until they joined or succumbed. By 1881 the Powder Trust was estimated to control 85 percent of the black powder market in the United States, and its position solidified in the next two decades.

The Gunpowder Trade Association was an exceptionally durable cartel in the volatile economic circumstances of the latter part of the nineteenth century. This durability can largely be attributed to the practice initiated by Henry du Pont, and emulated by Laflin and Rand, of buying into the other members of the association. As early as 1875 Du Pont had acquired control of Hazard, the third largest member of the cartel and also bought a 43 percent share of the largest West Coast black powder firm, California Powder Company, as well as numerous smaller fry. This strategy served two purposes: it gave Du Pont a larger share of the powder market indirectly through the allocation system; and it enabled Du Pont to examine the books of other companies as a further means of enforcing the sales agreements established by the trade association.[3]

Du Pont met the challenges of technological change in explosives in much the same fashion. In 1867 the Giant Powder Company, a California group, began manufacturing dynamite under a patent from Alfred Nobel, and established a branch in New Jersey, the Atlantic Giant Powder Company, three years later. Henry du Pont's initial reaction was total outrage and rejection: he pressured the Eastern railroads to stop carrying dynamite, arguing that it was unstable and dangerous. Lammot, however, prodded his uncle into a more constructive response. Du Pont's acquisition of the California Powder Co. in 1876 brought with it the process for a variant to Nobel's dynamite that was called "White Hercules," a mixture of nitroglycerine, sugar, and saltpeter, which had survived a Nobel patent infringement suit and was regarded as more efficient than the original dynamite.[4] In 1877 a new Hercules plant was established in Cleveland with Du Pont backing to meet the competition of Giant Powder in Eastern markets.

Three years later Lammot du Pont organized a joint dynamite venture with Laflin and Rand, the Repauno Chemical Company. By this time Du Pont and Laflin and Rand had also bought into Atlantic Giant Powder and in 1882 arranged an agreement for the cross-licensing of patents between Repauno and Giant, giving the black powder companies access to the Nobel patents. Two years later, however, Lammot happened to be visiting the Repauno plant in southern New Jersey when a workman allowed a vat of nitroglycerine to overheat. In the ensuing explosion, Lammot and others were killed. His fate recalled that of the original Pierre du Pont de Nemours, who had collapsed after fighting a fire in the Brandywine mills in 1817.

Lammot's loss was a considerable blow to the family, but the Repauno operation was continued under another Du Pont cousin, William. His successors, J. Amory Haskell and Hamilton M. Barksdale, a Du Pont in-law, instituted novel managerial practices to stimulate production and moved the dynamite firm directly into marketing and sales. Barksdale's approach to "systematic management," building on the ideas of Lammot du Pont, were to have a substantial impact on the reorganization of the Du Pont company after 1902. Meanwhile, in 1895, the Repauno plant and the Hercules and Giant properties in the East, together with twenty smaller dynamite ventures acquired by Du Pont and Laflin and Rand, were consolidated into the Eastern Dynamite Company, capitalized at $2 million.[5]

Du Pont also acquired a crucial foothold in the production of smokeless powder in the United States in this period, a step that would significantly affect later decisions to diversify into chemicals. Smokeless powder and its key ingredient, nitrocellulose, developed from the research of a Swiss chemist, August Schonbein of the University of Basel, in 1845. Schonbein was investigating the properties of cellulose, the woodlike fibers of cell walls of plants, particularly cotton cellulose. When treated with nitric acid and sulfuric acid the substance exploded with tremendous force. Schonbein carried out other experiments that would be of significance later in the cellulose field, but at the time his description of the explosive potential of nitrocellulose or "guncotton" caught the attention of military observers. Despite the dangers involved in research on guncotton British and French chemists experimented with the material over the next thirty years developing, among other things, cellulose plastics and filamented nitrocellulose, the first semisynthetic fiber. In 1886 Paul Marie Vielle of the Ecole Poytechnique was able to produce a diluted form of nitrocellulose explosive that retained the efficiency and power of Schonbein's guncotton but was more stable and could be controlled, tremendously extending the range and rate of artillery fire. Nobel also developed a compound of nitroglycerin and nitrocellulose called ballistite that was adopted as an alternative form of smokeless powder by Britain and Germany.[6]

In 1889 the U.S. Army's chief of ordnance approached Du Pont—the largest explosives supplier in the country—to recommend an investigation of smokeless powder. One of the youngest partners, Alfred I. du Pont, was dispatched to Europe, the first of numerous company officials to venture into foreign lands in search of new technology in the coming years. Rebuffed by the French, Alfred negotiated a licensing agreement with Coopal, a Belgian company. Dissatisfied with the results of tests with the Coopal process, Du Pont developed its own variant of smokeless powder at a research laboratory at Carney's Point, New Jersey, between 1890 and 1894. One of the junior research assistants at Carney's Point was Pierre du Pont, the eldest son of Lammot.

The processes for producing Du Pont's smokeless powder on a large scale had not been perfected when the Spanish-American War broke out in 1898. At the behest of the Army, Du Pont concentrated on producing another explosive, brown powder, by a process it had obtained under an 1889 license from the German company, Köln-Rottweiler. After that war, however, nitrocellulose smokeless powder was the basic military ammunition manufactured by Du

Pont, although black powder and dynamite remained the main source of income for Du Pont up to World War I. In 1914, when the company produced over 100 million pounds of dynamite and 5 million pounds of smokeless powder, dynamite production was down by over 70 million pounds from its high point in 1912 when Du Pont dynamite was blasting through Panama. Nevertheless smokeless powder was one of its most promising fields. It was a high profit item. Combined earnings for sporting and military smokeless powder contributed $1.3 million, roughly one quarter of the company's net earnings in 1914. By virtue of its cooperation with Army Ordnance, Du Pont held a monopoly of private smokeless powder production for the military from 1900 to 1916. Finally, the move into smokeless powder led Du Pont into expanded research efforts in the uses of nitrocellulose which proved to be the key to its diversification into chemicals during World War I.[7]

Corporate Reorganization, 1900–1907

Between 1870 and 1900 Du Pont became the largest explosives producer in the United States: sales increased by 2,500 percent in this period, compared with an average 155 percent increase for the industry as a whole. Assets grew at an average rate of 7.5 percent a year. Laflin and Rand, its closest competitor (and sometime partner) in the field, was only one sixth as large in terms of assets. Du Pont was not a "big business" like U.S. Steel, the billion-dollar behemoth created by merger in 1900, or Rockefeller's Standard Oil, whose net earnings in 1900 were nearly five times the total assets of Du Pont. Nevertheless, the Wilmington company was the dominant element in an important industry. Du Pont had risen to this position by skillful investments in other firms, acquisition of new technology, and a relentless pursuit of market stability. Although its field of endeavor widened and diversified in the twentieth century, the essential elements of Du Pont's approach to growth remained the same through World War II and virtually all of its expansion was financed from internal sources.

Du Pont at the turn of the century was in a troubled state. Lammot had been killed in 1884, and General Henry died in 1889, leaving no strong successor. A cousin, Eugene du Pont, took charge amid family bickering that grew stronger in the next decade. The younger family members, such as Alfred and Pierre, excluded from positions of leadership, were increasingly critical of the antiquated production methods prolonged by their elders. Labor problems emerged in the last decade of the nineteenth century, challenging the paternalistic tradition which was hard to maintain among the diverse operations that the company had acquired or established around the country.

The Du Pont dilemma was not just the familiar one of dynastic decline: it was a systemic problem, the direct result of rapid growth. The practice of buying into other firms had helped to maintain the cartel but also perpetuated inefficiency. There was little integration among the various branch plants, subsidiaries, and joint ventures. No one even knew the cumulative value of these assorted properties and investments. Except for individual operations such as Repauno, production facilities and methods were outdated. An appalled Pierre du Pont, recently graduated from the Massachusetts Institute of Technology in 1890, described the laboratory at the Brandywine mill as "deplorable . . . Equipment was almost nothing . . . No gas or electrical facilities, and a common kitchen sink and one ordinary three-quarter inch tap the water supply."[8] Underlying all these problems was the general lack of coordination among production units and between production, distribution, and sales operations. Du Pont was a burgeoning manufacturing body without a head, and a totally inadequate system of communication.

The executives of this ramshackle organization, all family members, were not totally unaware of the problems. The consolidation of dynamite operations under the umbrella of the Eastern Dynamite Company in 1895 was an attempt to improve coordination in that branch, but no further efforts were undertaken to extend this approach to black powder and smokeless powder. In 1899 Eugene arranged the dissolution of the partnership and the incorporation of E. I. du Pont de Nemours and Company in Delaware. The partners all received shares in the new company but Eugene continued General Henry's practice of one-man rule, and there was no attempt to rationalize the internal organization of the various properties. When Eugene died suddenly of pneumonia in January, 1902, the company faced a major crisis.

None of the senior family members on the board of directors was prepared to take over management of the firm, and they decided to sell out to Laflin and Rand. At this point, however, Alfred, the youngest board member, made a dramatic plea to buy the company from the others, asserting that "the business was mine by all rights of heritage . . . it was my birthright." Despite their lack of confidence in Alfred's business qualifications, the other family members reluctantly agreed to allow him first option on the purchase of the company. Alfred then persuaded two of his cousins, Thomas Coleman and Pierre, to join him in the bid.[9]

The three cousins were remarkably complementary in talents and background, although their personality differences would create serious friction in the company in later years. Alfred was a production man. He had negotiated for the smokeless powder patents in Europe and had directed the expansion of production in the Brandywine mill during the Spanish-American War. His experience outside production, however, was limited and his elders regarded him as reckless, although he later proved to be an astute speculator in Florida real estate and related ventures. By contrast, Pierre was grave, introspective, and systematic in his approach to his work. After a brief stint in the Brandywine mill, Pierre had gone to the Middle West to help manage Lorain Steel Company, a small mill left him by his father, Lammot. Under the tutelage of Arthur Moxham, Pierre mastered the fundamentals of cost accounting and also experienced firsthand the problems of a small expanding firm with insufficent working capital in a period of tight credit. Although his family investment diminished when Lorain Steel was absorbed by a New York-financed merger, the lessons of financial management were impressed upon him, reinforcing an inherent prudence and caution.[10]

Thomas Coleman du Pont was the most experienced and extroverted of the three cousins. Descended from a collateral branch of the family who had migrated to Kentucky in the nineteenth century, Coleman had worked his way up through the ranks of a coal mining company, joined briefly with Moxham, Pierre, and the businessman-politician Tom Johnson in the Lorain Steel venture, then struck out on his own in the electric railway field. By nature he was a salesman, relishing the horse-trading atmosphere of commercial negotiations.

Pierre, who examined the Du Pont books for Alfred, quickly recognized that the company assets, nominally $12 million, were undervalued by between $4 and $5 million because share holdings in other companies had not been reassessed at current market prices. He also grasped a significant second point: the combined shares of Du Pont and Laflin and Rand in other black powder and dynamite companies offered the prospect of control of more than 80 percent of the entire industry in the country. The three cousins decided to buy not only Du Pont but also Laflin and Rand and then consolidate the explosives field in a single operating unit. In short Du Pont was to establish in the American powder industry the kind of direct consolidation and coordination that Duisberg was vainly attempting to achieve in dyestuffs in Germany in this same period.

Thanks to Pierre's planning and Coleman's negotiating skills the feat was accomplished with a miniscule outlay of cash. The elder Du Pont shareholders were persuaded to exchange their voting shares for $12 million in thirty-year bonds issued by the reorganized company that was incorporated in March, 1902. The three cousins held 82 percent of the voting stock in the new company: Coleman retained 36 percent, Pierre 18 percent, and Alfred 28 percent (of which 10 percent represented his equity in the old company).

This task completed, Coleman and Pierre approached Laflin and Rand. Again Pierre concluded that the company's assets were undervalued, by $2 million owing to below-market evaluation of shares in other firms. Initially bidding for a 54 percent majority share of Laflin and Rand stock, the Du Pont cousins decided they needed to control all the shares in order to introduce the drastic reorganization contemplated. Two holding companies were established with a total authorized capital stock of $10.5 million, and $4.5 million in bonds of these companies, which were controlled by Du Pont, were offered to Laflin and Rand shareholders for their stock. In this complicated fashion the Du Pont cousins acquired control of the largest explosives companies in the country with a combined asset value of $17.5 million in 1903, for a total cash outlay of about $3,000.[11]

Three factors contributed to their success. First, the men at the helm of both the Du Pont and Laflin and Rand companies were for the most part aging and uninterested in managing their enterprises. They were ready for a takeover so long as their financial security was not jeopardized. Second, this was a period of rising expectations and intense merger activity in the American business community. Between 1897 and 1903 over 120 mergers took place in the United States absorbing almost 2,800 proprietorships, partnerships, and small family firms. Many of the largest mergers were initiated and financed by investment banking houses, but the spirit of consolidation affected other businesses with few ties to banks, including those in the chemical and explosives industries. As in the 1920s, and again in the 1960s, businessmen of this era regarded combination as the shortest and best route to expansion.

The third significant element was that the Du Pont cousins in their quest for consolidation were able to take advantage of a framework of mergers and joint ventures carefully established over many years by their predecessors. It is doubtful that they could have implemented a strategy of expansion by merger as rapidly if they had to rely on the resources of the Du Pont powder company alone. To do so would have required external financing and a significant dilution of family control of the firm; given Pierre's experience with Lorain Steel this course of action was improbable. The existing structure of overlapping investments in the explosives industry made the process of combination simple and financially painless.

By early 1903 the Du Ponts had begun to move to the next phase of consolidation and administrative reorganization of their domain, with the help of the Repauno veterans, Barksdale and Haskell. Virtually all the subsidiaries were absorbed into two new organizations: E. I. du Pont de Nemours Company of New Jersey, with a capitalization of $50 million, took over the black powder and dynamite companies, while Du Pont International Company of Delaware, capitalized at $10 million, took over the smokeless powder operations. Both of these companies were in turn controlled by E. I. du Pont de Nemours and Company of Delaware. Production facilities were concentrated in the largest and best located mills. The research laboratory at Repauno was expanded under Dr. Charles Reese and a new "experimental station" was established near the Brandywine mill, reflecting the importance Pierre du Pont and Arthur Moxham attached to continuous research.

Manufacturing was coordinated in three departments: black powder, high explosives (dynamite), and smokeless powder, under the general supervision of Alfred du Pont. The handling of raw materials and shipment of semiprocessed and finished goods was also centralized, as was the organization of engineering design and maintenance. To accommodate the new centralized staff Coleman du Pont bought a large office building in downtown Wilmington that has remained the corporate headquarters of the company to this day.

Marketing and sales were also brought under the direct control of the Du Pont staff. In March, 1904, the reorganized company withdrew from the Gunpowder Trade Association, ensuring its collapse. In place of the various sales agencies that had handled business for Hazard, Laflin and Rand, and the assorted other firms now absorbed into the new organization, Du Pont established a sales department under Charles Patterson and J. Amory Haskell, both from Repauno, where they had trained a sales force well versed in the technical details of explosives production and thus able to handle the specific demands of customers without delays. The department divided the country into geographical regions as well as product lines and established auxiliary staff to train salesmen, develop advertising, and maintain statistics for projecting future demand.

On the manufacturing side, the movement toward centralization had not drastically altered basic operations, since plant superintendents continued to handle routine matters. In sales, however, the change did have an immediate effect on the free-wheeling behavior of local agents and was the occasion for much grumbling and a few resignations, with significant results for Du Pont in at least one case. Despite these and other problems the transition was virtually complete by the end of 1904.[12]

At the peak of the new organization were three committees. The Administration Committee, consisting of the department heads, coordinated regular manufacturing and sales operations, which freed the Executive Committee, consisting of the three cousins, Haskell, Barksdale, and Moxham, to concentrate on "planning the future use of current facilities . . . and the development of new resources." In practice the Executive Committee took over many of the functions of the third committee, Finance, which represented the other family stockholders and in other companies acted as watchdog over management. At Du Pont the "triumvirate" dominated both committees.

Pierre and Coleman had broken with tradition by bringing outsiders into executive positions when they reorganized the firm and by establishing a stock bonus program as an incentive for managers. At the same time, however, they hoped to continue the "family firm" by encouraging younger family members to go into management and learn the business. Pierre was particularly attentive to the progress of his younger brothers, Irenee and Lammot Junior. In this effort, as in many other aspects, the family and company tradition of cautiously accommodating to new circumstances while preserving basic relationships was remarkably successful.[13]

The Antitrust Suit and Divestiture, 1907–11

One reason, although apparently not the most compelling one to Du Pont executives, for the abandonment of the Gunpowder Trade Association in 1904 was that it could be construed as a "conspiracy in restraint of trade" and thus illegal under the Sherman Act of 1890. Although that law had remained virtually unenforced against manufacturing for more than a decade after its enactment, Du Pont had been careful since about 1896 to avoid publicizing its cartel arrangements in the explosives field. When a more activist, or at least more vocal approach to the enforcement of the antitrust laws appeared under President Theodore Roosevelt in 1903 with the prosecution of the railroad merger, Northern Securities Company, the new heads of Du Pont took notice.

Du Pont was in the midst of reorganization in 1902 when the Northern Securities case made the headlines. Pierre du Pont and Moxham urged the rapid introduction of a centralized sales department and termination of the Powder Trust partly on the advice of the company's attorney, James Townsend. At this time the legal position under the Sherman Act of an integrated firm dominating a market, such as Rockefeller's Standard Oil, was uncertain, but the price-fixing practices of the Gunpowder Trade Association were clearly vulnerable to prosecution. Their view ultimately prevailed, although the financial and administrative benefits of consolidation were probably more persuasive to Du Pont's directors at this point.[14]

The reorganized Du Pont company, however, did not stop its practice of acquiring other companies and eliminating competition. While Pierre was in Wilmington planning administrative changes in 1903, Coleman du Pont went to California to strengthen Du Pont's position on the West Coast, leaving only one large competitor, Giant Powder Company, in the field. The company continued the policy of diversification by acquisition, buying up manufacturers of nitrocellulose-based explosives. The most important acquisition in this area was the International Smokeless Powder and Chemical Company which had engaged in brisk bidding against Du Pont for military contracts and had begun to branch into the manufacture of lacquers when it was taken over in 1904.

Du Pont also continued to absorb smaller operations in the black powder and dynamite fields. After dissolving the trade association and centralizing sales, Du Pont was in a position to reduce prices while increasing volume, measures that evoked cries of anguish from erstwhile cartel associates but which Haskell grimly defended as "survival of the fittest." Paradoxically the breakup of the Powder Trust had initially brought new competitors into the field, but very shortly they felt increasing pressure from Du Pont's price policies.[15]

One such competitor was Robert Waddell, founder of Buckeye Powder Company in 1903. Waddell had been a sales agent for Hazard, a Du Pont subsidiary, for twenty years. Like other agents accustomed to the camaraderie and bargaining over prices in the old Powder Trust, Waddell objected to the sales organization introduced by Haskell and Patterson. His motives in establishing his independent black powder firm were later a subject of much contention. Coleman du Pont believed Waddell's only aim was to nettle the big company until it bought him out, a not uncommon practice in the industry, as indeed did happen in 1907. Waddell asserted that he was driven out of business by Du Pont's deliberate price-cutting activities. Whatever the truth of the matter, Waddell's charges, which he made before the U.S. Senate Naval Appropriations committee in 1906, were widely publicized and given apparent substantiation by his submission of a large number of documents relating to the operations of the Powder Trust that he had accumulated before leaving Du Pont in 1904. This material formed the basis for an antitrust suit filed against Du Pont in July, 1907, the first of many such proceedings the company was to face in the ensuing half century.[16]

Du Pont was in a peculiarly disadvantageous position politically when the antitrust issue emerged. As a major supplier of military explosives Du Pont had always cultivated good relations with army and navy ordnance officers and had taken on experimental tasks such as the development of smokeless powder. Several members of the Du Pont family had political aspirations: "Colonel" Henry A. du Pont, son of "General" Henry, was elected to the Senate in 1906, assisted by Coleman who also had his eye on future public office as he tired of managing the powder company. These seemingly strong bastions of influence in Washington, however, proved vulnerable. The contacts with military bureaucrats were of little value in dealing with the White House and the Justice Department. Henry du Pont's appointment to the Senate Military Affairs Committee provided bad public relations for the company, lending credence to the claim that Du Pont interests were responsible for widespread political corruption, an image that was not helped by revelations about the usual unsavory bargaining that had preceded the senator's election by the Delaware legislature. Waddell's documents included references to a $70,000 Du Pont contribution to Theodore Roosevelt's reelection campaign in 1904, which may have forced the president's hand in pushing the antitrust suit. Against the emerging public pressure generated by Progressive era reformers the careful political fence-building of the Du Ponts proved to be quite rickety.

In the early stages of the imbroglio, Du Pont's executives were less concerned about the antitrust suit, which appeared to rest largely on practices discontinued after 1904, and more attentive to congressional proposals to expand the government's role in the production of smokeless powder. In 1906 Congress authorized construction of a small smokeless powder plant by the U.S. Army and an enlargement of the Navy's facilities at Indian Head, Maryland, that would provide a yardstick for determining future contracts with private suppliers. Two years later, in the wake of the antitrust suit, the House of Representatives amended a naval appropriation bill authorizing a further expansion of government smokeless powder production and enjoining military officials from contracting with private companies "acting in restraint of trade . . . or having a monopoly of the manufacture and supply of gunpowder in the United States." Du Pont had already closed down one of its three smokeless powder plants at this point and the threat of further cutbacks in government business contributed significantly to the company's decision to diversify beyond explosives.[17]

At this point Coleman du Pont effectively mobilized the company's political resources. A full-time lobby was established in Washington under the direction of Edmund Buckner, head of the company's military sales operations, and ordnance officers were induced to testify on behalf of Du Pont before Congress and the trial hearings. The military and naval appropriation bills went through Congress in 1909 with the restrictive amendment removed. Meanwhile, behind the scenes pressure was applied to the incoming Taft administration to drop the criminal proceedings that Roosevelt's attorney general had contemplated against Du Pont. Although the company lacked the public relations capability to counter the charges of Waddell and the Hearst press—defects that would be remedied in the future—Coleman displayed his formidable talents at back room manuevering.[18]

These moves proved to be at best rear-guard operations as the antitrust suit proceeded through the court. In June, 1911, the federal circuit court in Delaware handed down its decision, concluding that Du Pont had not only violated the Sherman Act up to 1904 but was continuing to do so. The circuit court based its judgment on two Supreme Court decisions released one month earlier, in the Standard Oil and American Tobacco cases. Although these precedents, by invoking the "rule of reason," would later form a shelter for large corporations against antitrust prosecution, the crucial point from du Pont's point of view was that market restrictions achieved through combination and consolidation by one enterprise were no more defensible than restrictive measures inaugurated by cartels. The legal rationale for consolidation expounded by Moxham and Pierre du Pont in 1903 collapsed as the court held that not only had the company continued to participate in the price-fixing arrangements of the Gunpowder Trade Association through 1904 but that its subsequent reorganization of sales operations aided "the combination in concentrating its power and fastening its hold on the monopoly which it had sedulously built up" since 1882.[19]

More unsettling to the Du Pont executives than the decision itself was the prospect of a court-ordered dissolution of the organization they had carefully constructed since 1902. Coleman du Pont was apparently so overwrought by the situation that he all but exhausted the political goodwill of the Taft administration by fruitless efforts throughout 1911 to persuade attorney general George Wickersham to reverse or ignore the court's directive for dissolution. At this point Pierre du Pont took over negotiations. It soon became apparent to him and attorney Townsend that the government, having achieved its symbolic victory over the trust, did not intend to institute a drastic reconstruction of the explosives industry.[20]

In the agreement that was ultimately negotiated and accepted by the circuit court in January, 1913, the Du Pont powder company was split into three parts, creating two new independent firms, Hercules Powder Company and Atlas Powder Company, which were to receive from Du Pont sufficient production facilities to supply 50 percent of the country's black powder market and 42 percent of the dynamite market. Military smokeless powder capacity was retained completely by Du Pont, thanks to the intervention of ordnance officials who insisted that this strategic material should be manufactured by one producer to ensure quality control. This arrangement proved to be crucially important to the future of Du Pont. Within less than two years the company was to become a major supplier of munitions to the Allies as well as to the United States government in World War I. The tremendous earnings accumulated from these sales would finance Du Pont's diversification into chemicals. Atlas did not develop a significant smokeless powder capacity. Hercules entered the field in 1916 by acquiring Union Powder Company which had a nitrocellulose plant, but was producing only about one tenth the capacity of Du Pont in this field in 1918.[21]

In retrospect the dissolution order did little to hamper the growth of Du Pont, and in the judgment of some observers, also did little to reduce the company's domination of the explosives industry in the United States. Although Du Pont's total assets were reduced by $20 million (about 33 percent) as a result of dissolution, most of the loss was in increasingly obsolete black powder production facilities. Du Pont shareholders, family members for the most part, received $10 million in bonds and $10 million in common stock from Hercules and Atlas. Du Pont shareholders allegedly retained control of the two companies even though 50 percent of the common stock had been deprived of voting rights by the court order. The headquarters of the two new companies remained in Wilmington.[22]

Nevertheless, the experience of the antitrust suit and dissolution decree had a traumatic and lasting impact on Du Pont's chief executives. Pierre's younger brother, Lammot du Pont, observed later than the decision "served notice to the Du Pont company that it could not expand in the explosives field . . . and that was a very powerful influence in branching out into other lines."[23] Although historians of the company have argued that there were other more compelling considerations, the antitrust suit obviously did ultimately reinforce arguments for diversification advanced on other grounds.

More important, the events of 1911–13 had a permanent influence on the legal and political outlook of Du Pont's leaders. One immediate effect, discussed below, was the renegotiation of agreements with foreign explosives producers to avoid further antitrust difficulties. Du Pont representatives over the next three decades would insist on legal usages in international arrangements that could not be construed as market restrictions, to the frequent bemusement of their British and European counterparts.

Finally, the frustration of counteracting a legal onslaught that had widespread public support produced among Du Pont's chief executives a profound distrust of popular politics coupled with a determination to organize the company's public relations to create a more favorable image, much as the Rockefellers were undertaking at this same time. The task was never entirely successful: the internal squabbles of Du Pont family members received much lurid publicity over the next few years and the heads of this large private corporate domain could not always conceal their basic contempt for politicians and public opinion. There was, however, a continuing effort to combine behind-the-scenes maneuvering with bureaucrats and political party leaders with energetic publicity on behalf of Du Pont as the creator of "better things for better living" for society as a whole. The fruits of these efforts were to be reaped in the 1920s but the political power and economic prominence Du Pont acquired in that decade may also ironically have contributed to its legal difficulties in the ensuing years.

CHAPTER THREE

Du Pont and the International Explosives Industry

While Du Pont was carefully constructing the Powder Trust to control competition in the American market, the European explosives producers faced similar problems of overcapacity and competition. By the 1880s the European continent could not absorb the increased production spawned by improvements in the efficiency of plant output and an intense rivalry began to develop in peripheral markets around the world. The solution, embraced within national boundaries for years, was to establish mechanisms for regulating the industry at the international level. Instead of cutthroat competition, European powder makers began to negotiate a series of international agreements to allocate market territory, control production, fix prices, and protect individual companies that entered these cartels. By the end of the century the European manufacturers had assumed a decidedly cosmopolitan outlook toward the explosives trade, emphasizing "international markets, international competition, international organization. What went on at home . . . was of secondary importance."[1]

Across the Atlantic, however, the American explosives manufacturers had a more parochial outlook. Du Pont, like its competitors and partners in the Powder Trust, was primarily concerned with conditions in the United States and up to the 1880s did not perceive any real threat from overseas. A decade earlier, Europeans who ventured into this market were repelled by the chaotic state of competition with the continuous entry of smaller firms and the intense rivalry of larger ones such as Du Pont and Laflin and Rand.

Between 1880 and 1886 Alfred Nobel, the inventor of dynamite and the dominant figure in the European explosives trade, reorganized the various quasi-independent explosives producers in Britain, Germany, and elsewhere into the Nobel Dynamite Trust, and proceeded to launch a new foray into the American market through a small New York firm, Standard Explosives Company. Over the next two years the Nobel group and the American Powder Trust engaged in a contest of wills, threatening to invade each other's hitherto sacrosanct domains. By 1888, however, both sides recognized the wisdom of collaboration and negotiated an "American Convention" that would reserve the U.S. market for Du Pont and its partners in the Trust in exchange for a promise to keep out of the traditional European markets which embraced most of Asia, Africa, and Australia as well as the Continent. Equal rights were awarded to both sides of the convention for Canada, Mexico, Central and South America, and the Caribbean Islands as well as Japan, China, and Korea.[2]

The 1888 agreement between the Americans and the Anglo-German group was only the first of a series of international agreements that continued well into the twentieth century. Each agreement necessitated negotiating sessions among the parties which gradually drew the Americans closer to their European counterparts. This increased contact with business in the British Isles and on the Continent had wide ranging implications. Most importantly, it forced the American manufacturers to acquire a heightened awareness of Europe's technological and, in some instances, organizational superiority. Du Pont's entry into the international scene took place in the context of an established system of cartels that not only buttressed her position at home but also extended the promise of future access to technical innovations in explosives and eventually in the chemical industry as a whole.

Du Pont's International Diplomacy, 1897–1907

Despite the initial stability produced by the 1888 convention, the threat of overproduction continued to loom over producers on both sides of the Atlantic. The development of smokeless powder for military propellants introduced a new factor into the situation, and Du Pont hastened to acquire and develop its own variant in this field, meanwhile reorganizing the Eastern Dynamite Company, which was intended in part to carry out export operations for the Powder Trust should relations with the Europeans break down. When a German firm, Rheinisch-Westphalische Sprengstoff, proposed in 1897 to open an explosives operation in New Jersey, in Du Pont's own back yard, the Americans determined that the time had come for a new agreement.[3]

After several months of negotiations, a new convention was drafted covering smokeless powder as well as blasting powders. More elaborate than the first agreement, this one provided for the usual allocation of home markets to each main party, and for the pooling of profits made from sales by either side in "syndicated territory" that included parts of South America and the Caribbean. In case either party received orders for military smokeless powder from the other party's government, a royalty would be paid by the foreign seller. For the most part, this agreement eliminated the competitive marketing problems of the 1890s and laid the groundwork for a gradually expanding intimacy between Du Pont and the European firms. Since most of the negotiations were carried out by British representatives, the relationship between Du Pont and British Nobel was to become particularly close.[4]

Up to the beginning of the twentieth century, Du Pont's business diplomacy was in large measure limited to defensive gestures, to block potential European rivals from entering the U.S. market. The consolidation of the company in 1902, however, brought with it a new phase in Du Pont's foreign relations. For the first time the company began to contemplate a more venturesome approach to overseas business. Consolidation itself ensured Du Pont a firmer grip on the American market which translated into increased leverage vis-à-vis the Europeans. In addition, the establishment of a Development Department with responsibility for long term strategy increased Du Pont's awareness of the international nature of its operations.

In 1903, for example, in the process of reviewing the company's flow of raw material supplies, the Development Department concluded that the company was vulnerable to interruptions in its supply of nitrates from Chile. The Department recommended the purchase of nitrate lands in Chile to ensure its independence of outside suppliers, mostly Germans and British. Although these suggestions were not acted upon immediately, a year later Pierre du Pont included a trip to South America on the itinerary of one of his fact-finding missions abroad.[5]

Pierre du Pont's interest in this subject derived not from a strong desire to expand the company overseas but rather from his position as financial arbiter for the allocation of new investments. His cousin Coleman, however, soon developed an enthusiastic interest in foreign matters which led to the formulation of an ambitious international strategy for the firm. Coleman's initial foray into the world of foreign investment came about as a result of problems Du Pont encountered in Mexico.

Mexico had been designated "American" territory under the 1897 agreement and was theoretically off-limits to European competitors of Du Pont. Around 1901, however, the "Latin group" of explosives producers—a small body of independent explosives manufacturers who operated outside the Nobel umbrella—began to invade Mexican markets. In 1904 Coleman met in St.

Louis with Siegfried Singer, one of the heads of the "Latin group." The two men immediately established a rapport, and Coleman concluded that the Mexican problem could be resolved by Du Pont's acquiring majority shares in the "Latin group" which incidentally would strengthen Du Pont's position in relation to the Anglo-German Nobel group.

This proposal was only the beginning for a more grandiose plan for a world trust in explosives that Coleman unveiled to Henry de Mosenthal of the Nobel group on a trip to London in 1906. Apparently Coleman did not clear this proposition with his partners in Wilmington and the astonished British Nobel people chose not to take it seriously. Coleman's continuing pressure to absorb the "Latin group" led to a confrontation with Pierre later that year.

Pierre's objections to Coleman's strategy stemmed primarily from his conviction that Du Pont's most lucrative market continued to be in the United States. But he had also investigated the financial conditions of the "Latin group" during his 1904 tour abroad and concluded that a Du Pont purchase of a majority share of Singer's organization, far from strengthening the American company, might tie a millstone around it by adding the burden of a large, unwieldy, and basically unsound network of distributors. Furthermore, Pierre opposed what he regarded as unnecessary provocation of the major European producers. Cooperation had worked at home and there was no reason to doubt its efficacy abroad. By 1906 Pierre's cautious approach had persuaded other members of the Executive Committee, and Coleman's proposal was vetoed.[6]

Pierre du Pont's philosophy regarding a cooperative approach to European producers became deeply embedded in the Du Pont company's later involvement in the international chemical industry. As Du Pont began to expand its product lines beyond explosives it continued to apply these principles, which in fact came to be regarded as essential to the success of diversification, by ensuring Du Pont access to foreign technology. Whenever opportunities to expand abroad arose, Du Pont executives repeated Pierre's pronouncements about the primacy of the domestic market, the advantages gained from industrial cooperation, and the necessity for securing advanced technology. This point of view received a classic formulation in a 1916 Executive Committee memorandum that counseled Du Pont to "refrain from active business in foreign countries unless a foothold is gained with those already engaged in business. . . . Otherwise we would become known as 'invaders' of the business of others, our business would be less profitable and we would have less opportunity of obtaining information from foreign sources."[7]

The debate within Du Pont in 1906 over the acquisition of the "Latin group" had wider ramifications. The Nobel group was concerned over the incursions by Singer's organization into Mexico and feared that a similar situation might develop in Canada where railroad construction and mining was stimulating increased demand for explosives. Equally troubling was the status of Du Pont's foreign agreements in view of the threat of antitrust proceedings that loomed over that company. In 1905 J. A. Haskell of Du Pont warned the British that the legality of the 1897 agreement was in question. British Nobel was particularly alarmed by this development which appeared to jeopardize the cartel. In 1906 the Nobel group held a meeting at Cologne to try to resolve these difficulties.

Du Pont's decision not to buy the Singer companies helped settle one of the problems. The Americans' interest in European technology, not only in explosives but also in related cellulose-based fields paved the way for an agreement that appeared to circumvent the antitrust question. The new agreement negotiated in 1907 eliminated references to territorial divisions in markets. To replace these provisions Du Pont and the Nobel group arranged for a comprehensive exchange of patents and processes that included restrictions on the licenses to specific territories paralleling the marketing territories that had existed under the 1897 agreement. In this manner the companies could continue to regulate the world explosives industry and Du Pont hoped to avoid charges under the Sherman Act. At this point the potential benefits of such arrangements as a mechanism for acquiring foreign technology were not fully recognized by Du Pont, but that dimension was to become increasingly important in the future.

Anglo-American Links, 1907–14

Although Coleman du Pont's dream of a world trust in the explosives industry never came to pass, Du Pont's contacts with the European companies were solidified by the 1907 agreement and led to even more intimate relations over the next few years. Significantly, these contacts were made primarily with the British representatives of the Nobel group and the groundwork was laid for bilateral agreements that initially only complemented the general Anglo-American-German understanding, but would eventually replace them.

While the British were negotiating on behalf of the entire Nobel group in 1897 and 1907, by the time of the second general agreement they had concluded that amicable relations with Du Pont might be more beneficial to the British Nobel companies than to the Germans. Certain individuals in the British Nobel organization, particularly de Mosenthal and Harry McGowan, were especially convinced on this point. Moreover, during the course of negotiations over the renewal of the 1897 agreement, the British group became aware of several areas of mutual concern to themselves and Du Pont, exclusive of the German group.

A first step toward closer bilateral contacts was taken in 1908 when the British Nobel company renewed contacts with Singer's "Latin group" that led to an arrangement that eliminated competition with the Americans in Mexico. At the same time Nobel offered Du Pont an interest in a Canadian firm, Hamilton Powder Company. This small interest provided the two parties with an opportunity to work together on a joint venture and thus develop a better understanding of their respective organizations and methods of doing business. Observers sent by British Nobel to Wilmington in this period sent back glowing reports about the efficiency and innovative nature of Du Pont's financial and manufacturing procedures. In 1909, McGowan, who was emerging as the leading figure in the British Nobel firm, began negotiations in the United States to bring the two companies even closer together in the lucrative Canadian market. In 1911 Nobel merged Hamilton with three other small Canadian firms, including a cartridge manufacturer and a chemical producer, and formed a new company, Canadian Explosives Ltd. (C.X.L.), in which Du Pont was offered a 45 percent share. Within several years after its founding C.X.L. proved to be an enormous success.[8]

The Canadian joint venture was the key to future relations between Du Pont and the British. As in most corporate partnerships a major factor in its success was the establishment of close personal ties and an aura of mutual trust between individuals in Wilmington and London. McGowan in particular cultivated relations with Du Pont executives and would become the chief architect of the Du Pont–Nobel alliance that expanded into the chemical field during the 1920s. His confidence in Du Pont's industrial prowess and desire to establish an enduring basis for the Du Pont–Nobel relationship converged with the conviction among Du Pont's heads that their interests in protecting the domestic market could be best served through collaborative efforts abroad.

The 1907 agreement lasted for only six years. Ironically, despite the pains taken by all parties to accede to Du Pont's requests for the delicate handling of the antitrust problem, the cancellation of the agreement in 1913 was precipitated by Du Pont. The divestiture proceedings had persuaded Pierre du Pont that the patent exchange arrangement provided insufficient protection against further legal assaults on Du Pont. Eventually, Pierre was convinced that the long-term benefits of an international agreement outweighed the dangers of antitrust prosecution at home, but he insisted that a new patent exchange must include a specific royalty provision in order to counter charges that the patent agreement was simply a mask for a cartel. Although Nobel's negotiators considered this demand a form of nit-picking that might undermine the broader relationship they were seeking, Pierre's position was accepted. Under the new agreement, concluded on July 2, 1914, provisions were included for royalties on patents, licenses, and equipment.[9]

A month later war broke out in Europe. While the British took the position that the agreement should remain in place, Pierre argued that with Britain and Germany on opposite sides, enforcement of the terms would be impractical. Within a few months the Allied governments were approaching Du Pont with contracts for military supplies, and the Americans' unilateral suspension of the agreement ensured that they could fill these orders without encountering charges of bad faith. The explosives cartel thus fell, a victim of both the war and American antitrust laws, just as the chemical agreement was to collapse under similar pressures during World War II. When relations among the explosives producers were restored in the aftermath of the conflict, the initial links were made between Du Pont and the British, a circumstance that would have a lasting effect on the shape of international relations in both the explosives and chemical industries over the subsequent decades.[10]

CHAPTER FOUR

The Impact of War, 1914–18

World War I has been called the "chemists' war" much as World War II was to be the "physicists' war."[1] For the first time in modern history the military significance of science and the strategic importance of chemicals as raw materials and catalysts for incendiary weapons was recognized and exploited by the belligerent nations. As a result the full weight of government support was placed behind the rapid development of technology and industrial capacity in chemicals. The war thus constituted an incubator not only for technical devlopment in such areas as cellulose-based explosives, poison gas, and synthetic ammonia, but also fostered the growth of large-scale industrial and financial amalgamation in chemicals in Europe and the United States.

The warring nations each had their own particular demands on their domestic chemical industries. Among the Allies, including the United States, rapid development of dyestuff production was promoted, not only because dyes were important for the continuing operation of their domestic textile industries, but also because intermediates extracted from coal tars, in particular phenol and benzol, were crucial ingredients in the highly explosive shell charges, picric acid (trinitrophenol) and T.N.T. (trinitrotoluol). The absence of a strong domestic dye industry was attributed to the sinister designs of the Germans, and appropriate changes in tariff and patent laws were introduced to encourage growth. In Germany, dyes and intermediates were abundant, but the Allied blockade cut off supplies of another critical material, Chilean nitrates, the ultimate source of saltpeter used in black powder and, more important for military purposes, nitric acid for high explosives. To overcome this problem the German government provided financial aid to the chemical companies that had been cautiously developing processes for the fixation of nitrogen from the atmosphere since about 1910.

The results of these war-borne programs of development were dramatic. In Britain domestic dyestuffs production increased from 4,000 tons in 1913 to 22,500 tons by 1920. In the United States the transformation was even more remarkable. A minute organic chemical industry consisting of seven firms employing 500 people, with a total capitalization of $3 million, and producing 3,000 tons of dyes in 1914 burgeoned into a highly visible element in American chemicals, producing over 29,000 tons of dyes by 1920, exporting about 10 percent of this total, and attracting more than one hundred and twenty-eight companies into the field, including the two largest, Du Pont and Allied Chemical and Dye Company, whose earnings from dye sales in 1920 equaled almost four times that of the entire American dye industry before the war. Meanwhile in Germany the production of synthetic nitrogen soared to 185,000 tons by the end of the war, 50 percent of it from the atmospheric fixation of nitrogen, a process that was not even commercially operating before the war began.[2]

The war also generated an unprecedented diffusion of chemical technology, particularly when the Allied governments seized enemy properties and patents and turned them over to domestic chemical manufacturers. The significance of these drastic measures, exceptional for countries normally committed to the protection of property rights, has been exaggerated; but there can be little doubt that valuable information was made available to a much wider commercial and technical audience than had hitherto been the case, and British and American manufacturers exploited the situation. Furthermore, these dramatic events made the chemical industry front-page news, arousing the interest of new investors at a time when capital requirements were outpacing the internal resources of manufacturers. The big chemical combinations of the 1920s—Allied Chemical and Dye in America, Imperial Chemical Industries in Britain, Montecatini in Italy, I. G. Farben in Germany—were all at least in some measure the products of the wartime dissemination of technology in dyestuffs and synthetic nitrogen, the promotional activities of wartime governments, and the concurrent public awareness of the military significance and commercial potentialities of chemicals.

Du Pont participated in these developments. The Wilmington company entered the dyestuffs field as early as 1916 and later made an equally substantial plunge into synthetic ammonia. At the same time, however, Du Pont's course was shaped by internal considerations and the industrial world view of its executives. Du Pont's path to diversification varied in some respects from the direction taken by other large chemical firms, at home and abroad. Building on a base of nitrocellulose production, Du Pont investment flowed into solvents and paints, into exotic areas such as celluloid film and artificial silk, and ultimately into the seemingly improbable field of automobile manufacturing. Diversification decisions were based on a careful assessment of financial as well as technical factors, and, with a few notable exceptions, were little influenced by contemporary fashion in investment markets. To a large extent this imperviousness to outside influence derived from Du Pont's singularly strong position as a company that could generate most of needed investment capital from earnings. Here again the impact of the war was critical. Without the vast influx of "war profits" accruing to Du Pont as the largest supplier of smokeless powder and high explosives to the Allies in 1914–18, the company's move into—and beyond—chemicals would have been much smaller, more gradual, and ultimately probably less significant for the industry as a whole. Fortune smiled upon Du Pont, and the company's executives wisely chose to exploit the opportunity to ensure that in the future they need not rely on the whims of war and the vagaries of politicians and military bureaucrats.

The Offspring of War

In 1913 Britain imported 20,000 tons of dyes valued at $9 million, supplying over 80 percent of her market. Most of the imports came from Germany and Switzerland. This situation was not in itself regarded as cause for alarm in peace-time since Britain also exported acids and soda more than the value of dye imports, and retained domination of the textile markets of the Continent. The prospect of war and military dependence on dye intermediates for explosives, however, cast a different light on the situation. During the Boer War Britain's reliance on imports of benzol and phenol from a potentially hostile Germany became a matter of public concern, and in 1907 the patent law was revised to require foreign companies to establish manufacturing facilities in Britain within four years in order to retain their patents. In 1911 the Swiss firm, Ciba, acquired a British subsidiary to comply with the law, and the German dye makers, Hoechst, built a British plant to produce synthetic indigo as well as the drugs salvarsan and novocaine.

Nevertheless, the British chemical industry was largely unprepared for war in 1914. The largest domestic dye firms, Read Holliday and Levinsteins, produced a limited range of dyes and intermediates, and neither the private companies nor the government had thought to stockpile supplies. For a time the British had to import German dyes through neutral countries, a situation exploited by the Germans to acquire rubber and cotton for their own war needs. In 1915, however, the British government embarked on a program to encourage domestic dyestuffs. German-owned patents were made available to British dye makers and the Hoechst indigo plant was confiscated, and later acquired by Levinstein. The smaller dye company, Read Holliday, was reorganized with $7 million capitalization in 4 percent debentures put up by the government. The major peacetime dyestuff users, the large textile makers' trade associations, were pressured into buying shares in the new company, christened British Dyes Ltd., to increase total capital investment to $10 million. In addition to supplying the government with intermediates for explosives, British Dyes was expected to expand the range of domestic dyes so that Britain "would not exchange her dependence on one foreign country for dependence upon another," an oblique reference to the Swiss. To this end the new company sought to merge all British dye manufacturers, including Levinstein, under its corporate umbrella.

Although these emergency measures were successful in expanding wartime domestic production, the longer range aims of the program proved harder to attain. Textile manufacturers blocked the imposition of a protective tariff on dyes for the duration of the war and clearly preferred the more varied and higher quality Swiss dyestuffs. Furthermore, the planned amalgamation of British dye makers was delayed until 1919 due to bickering between Levinstein and British Dyes Ltd., which was to continue even after the merger which established a $50 million holding company, and would weaken it in meeting postwar competition not only from the Germans and Swiss but also from the emerging American dye industry.[3]

In the United States the dye industry before the war had been even smaller than in Britain. The second largest consumer of dyes in the world, the United States imported 90 percent of her needs in 1913. Of the handful of domestic producers, only two also manufactured their own intermediates: Schoellkopf's Contact Process Company and Benzol Products, a small joint venture established in 1910 by General Chemical Company, the large sulfuric acid maker, and Semet Solvay, the coke and coal tar supplier to Solvay Process Company.

As noted earlier, American patent laws, which lacked domestic manufacturing requirements, and American tariffs on dyes were largely responsible for this state of affairs. Since 1883 duties on imported dyes and intermediates had been steadily reduced, largely owing to the persistent lobbying of textile manufacturers who had a foothold in both major political parties from their New England Republican and Southern Democratic bases. Ironically, while the giant steel industry and the textile industry itself received the full benefits of tariff protection, organic chemicals, a genuine "infant," languished.[4]

Following the outbreak of war in Europe in 1914 the German government sought to acquire supplies of American cotton and wool in exchange for dyes and medicines, and for a short time the British, who were also procuring needed dyes on this barter basis, allowed a continued flow of German exports to the United States. By early 1915, however, the British position hardened and a "coal tar famine" ensued, affecting not only the textile industry but also leather goods, paints, furniture, and printing. Although rising prices for dyes attracted many smaller firms, including Monsanto and Hooker Chemical, into the field, they were handicapped by a lack of capital, an unfamiliarity with the technology that contributed to numerous fires and accidents, and above all an inadequate supply of intermediates. Technically capable producers, such as Schoellkopf, were reluctant to expand as long as the postwar prospects for the industry remained uncertain; and these doubts were shared by potential investors in the field. Many entrepreneurs who did jump on the bandwagon, including Antoine du Pont, Coleman's younger brother, shortly went under as prices began to level off and new sources of capital remained unavailable.

As early as November, 1914, the American Chemical Society advocated a major revision of dyestuff duties, and in December, 1915, a bill was introduced by Representative Hill, a Connecticut Republican, that would impose a per-pound duty as well as *ad valorem* rates on dyes and intermediates, and put coal tar itself on the free list. The bill was defeated, but a second effort was mounted with President Woodrow Wilson's support by Senator Claude Kitchin of North Carolina in 1916. Wilson evidently regarded the bill as a war measure, but news of the amalgamation of German dye firms into a single giant organization added urgency to the protectionist pleas of American organic chemical manufacturers. In its final form the Kitchin bill incorporated the increased duties on dyes and intermediates but exempted certain dyes, including the azo group and indigo, from the per-pound surcharge. In keeping with the general free-trade beliefs of Wilson and other Democrats, the bill also provided for the gradual reduction of the duties and their ultimate termination if American producers could not supply 60 percent of the domestic market by 1921.[5]

The Kitchin bill or Emergency Dye Tariff Act of 1916 thus constituted a sword of Damocles over American dye makers, but it also provided the foundation for an expanded industrial base. The U.S. Tariff Commission, which was directed to review the development of coal tar chemicals to ensure that the 1921 provision would be adhered to, became a proponent of the American dye industry and provided a clearinghouse for information for the emerging trade association, the American Dye Institute and its successor, the Synthetic Organic Chemicals Manufacturers Association. The War Industries Board also promoted cooperation in chemicals, as in other industries, and military demand for benzene, phenol, and toluene, boosted the market for coal tar intermediates, encouraging vertical integration among dye makers.

Wartime demand and the prospects for a reasonably stable postwar market for dyes encouraged the entry of bigger firms and the infusion of larger amounts of capital from the investing public. In May, 1917, National Aniline and Chemical Company was established with assets of $9 million, a total capitalization of $20 million, and aggregate dye output of 17,500 tons per year. This first large-scale combination in chemicals was the result of a merger of Schoellkopf's company, which specialized in cotton dyes, W. G. Beckers Company, a much smaller New York operation specializing in wool dyes, and Benzol Products, which supplied aniline oil to both dye makers. The merger was orchestrated by Eugene Meyer, a New York financier and promoter who had supported the Beckers company in 1914 when it was "a small operation on the second floor of a Brooklyn garage." Although the dominant interests in amalgamation initially were Schoellkopf and William Nichols of General Chemical Co., Meyer brought in an industrial engineer and long-time associate, Orlando Weber, to represent his interests. After a period of internal squabbling, Meyer's faction ousted Schoellkopf, and Weber took over the presidency of National Aniline. Weber, a hard-driving temperamental man, son of a labor organizer, had already resurrected the Maxwell Auto Company for Meyer when he took on this new task. The company was reorganized, concentrating dyestuff production at Buffalo, New York, and intermediates at Marcus Hook, Pennsylvania. Centralized sales and research units were set up, similar to those at Du Pont. By 1918 National Aniline's total assets had increased to $62 million.[6]

Du Pont also made public its plans to enter the dyestuff field at this time. The move had been contemplated for a long period beforehand. There was a certain logic to the explosives company's diversifying in this direction. The Repauno plant had long produced certain intermediates such as anthracene and diphenylamine, and internal production of these chemicals increased with the rising wartime demand for explosives. The prospect of postwar inventories of these intermediates was not alluring.

Balanced against these considerations were Du Pont's lack of technical capability in the actual production of dyes from intermediates, and the probable need to build new dye facilities which might have little value if the dye market collapsed after the war. Pierre du Pont and his younger brothers, Irenee and Lammot, were inclined to override these objections. As Lammot, who headed the dye division, later maintained, "Du Pont had developed a large organization of chemists, chemical engineers and chemical operators devoted to war production . . . The organization was all set for some venture in the chemical industry and . . . coal tar chemicals were a shining mark for those who had the knowledge, ability, and courage. The Du Pont company felt it had these although it did not have a specialized knowledge of dyestuff production or application."[7]

Despite this enthusiasm at the top, the Du Pont executives remained cautious until Congress passed the Kitchin bill in September, 1916. At this point the company acquired the services of an exceptionally talented dyestuff salesman, Morris R. Poucher, who had handled B.A.S.F.'s dye sales in the United States before the war. Poucher was to prove to be equally skilled as a lobbyist for Du Pont and the dye industry in years to come. Partly upon Poucher's recommendation, Du Pont decided to concentrate efforts on the manufacture of dyes which had not been attempted by other American companies for technical reasons, in particular the "vat dyes" derived from anthracene whose adhesive qualities on cloth were exceptional and were made almost exclusively by the Germans; and indigo, which was in demand in China, the largest dye market in the world, as well as the United States.[8]

A Du Pont group consisting of Poucher, Dr. Arthur Chambers of the company's chemical department who had initiated research on dye intermediates, and J. Amory Haskell, head of the sales department who had some negotiating experience with the British, went to London. Their object was to procure a technical agreement that would give Du Pont access to new dye processes.

The Du Pont emissaries arrived in Britain when the dye manufacturers there were embroiled in ultimately futile negotiations to merge British Dyes and Levinstein. Haskell rapidly concluded that British Dyes was a hopeless case, although Harry McGowan of Nobel, who was involved in the dye merger talks, proposed a joint venture with Du Pont that would give the two explosives firms control of the British dye industry. Levinstein appeared more attractive, however, for it was a well integrated company which had recently taken over the Hoechst indigo works in Britain and thus had direct access to crucial German dye processes. At the end of a month of bargaining, Levinstein agreed to give Du Pont technicians access to the indigo plant and to license its patents in the United States for royalties of $1.25 million over ten years. The two companies also agreed to share future patents and processes and to cooperate in American dye sales since Levinstein already had a sales agency there. These last two provisions proved unworkable in practice and were canceled in 1921, but in the meantime relations deteriorated to the point where Levinstein's American agent sued Du Pont and the whole Levinstein contract became a political issue in the debates over the dye tariff. For the moment, however, Du Pont had its basic requirement, access to dye technology. In February, 1917, the explosives company began building a dye works at Deepwater, New Jersey, and subsequently acquired a small firm, United Piece Dye Works. Its total investment in dyes between 1917 and 1921, exclusive of the Levinstein royalty, was $29 million.[9]

The Du Pont venture into dyestuffs was in certain respects unwise. The protective tariff and the diffusion of dye technology brought many small competitors into the field, joined after 1921 by the resurgent German dye firms, now combined into a centralized export sales organization under I. G. Farben. The venture into the Chinese dye market urged by Poucher proved to be a costly fiasco, as will be related later. Pierre du Pont later acknowledged that the company lost more than $40 million on dyes before it began to show a profit, although he may have been exaggerating for effect since the statement was made in the context of an antitrust suit. Nevertheless, the losses were substantial and would certainly have crippled a smaller, less diversified firm or one more dependent on external financing. The dyestuff enterprise also obliged Du Pont to mount its largest political lobbying effort since 1908, to ensure the preservation of a protective tariff, providing further notoriety for a company already stigmatized as a "merchant of death."

At the same time, however, Du Pont reaped the ultimate benefits of persistence in this venture, acquiring a dominant position in the American dye market. The floundering textile industry became something of a dead weight for dye producers by the 1930s, but by this time Du Pont occupied a solid one third of the dye market, and by 1950 its gross annual sales in this field were running at $20 million more per year than its total investment in dyes over the first twenty years of development. The company took pride in its patriotic role in freeing the American dye industry from the foreign yoke. Perhaps equally important, the dyestuffs venture gave Du Pont a foothold in the synthetic organic chemicals field which was to prove increasingly important to the development of the company and the industry in the ensuing decades as du Pont developed, among other things, plastics, food preservatives, and, especially, synthetic fibers.

The rapid development of the ammonia synthesis process for the fixation of atmospheric nitrogen on a commercial scale was one of the principal achievements of the chemical industry in World War I. Various methods for nitrogen fixation had been developed in the decade before the war, but without the pressure of military demand for the chemical, the heavy capital investment costs of the process would have delayed the introduction of large-scale production in this field for a long time. As it happened, however, the warring states helped underwrite development costs, attracting private investors to the field and contributing to the emergence of large, heavily capitalized businesses in chemicals. In a more indirect fashion, the financial and technological resources applied to the creation of the synthetic ammonia industry laid the groundwork for research and development in synthetic fuels, synthetic rubber, and the whole field of petrochemicals.

Nitrogen in its acid form is an essential ingredient in most explosives: sodium nitrate in black powder, nitroglycerin in dynamite, nitrocellulose in smokeless powder. The military need for this material is apparent, but the research that led to the discovery of the ammonia synthesis was initiated for reasons far removed from considerations of war.

In 1898 Sir William Crookes, an eminent British scientist, predicted worldwide famine by the early 1920s because of the declining productivity of arable land. This forecast, which has a familiar ring to audiences today, was predicated on the assumption that the natural replenishment of soil fertility through the absorption of nitrogen by plants could not keep pace with the current rate of agricultural production. The crisis would come when supplies of sodium nitrate from Chile as a source of fertilizer were exhausted. Other forms of fertilizer could not ensure continued high yields of grains, and the other major source of nitrogen, sulphate of ammonia from coke ovens, could recover only a fraction of what was needed for fertilizers, not to mention explosives, dyes, and refrigerants.

Crookes's warnings received much publicity and had a substantial, although perhaps exaggerated, impact on subsequent events. As Crookes himself went on to observe, over 80 percent of the earth's atmosphere consists of nitrogen which would provide an inexhaustible supply of the material if it could be converted or "fixed" by combining it with other substances usable as fertilizers. Developments in physical chemistry and electrical engineering had already progressed to the point where nitrogen as a gas could be drawn from the atmosphere, and scientists were experimenting with methods of producing nitric acids through electrolytic processes.[10]

One method, the "arc process," had been examined by Crookes and others in the 1890s, and appeared to be the most promising despite the fact that it required large amounts of electric power, as the conversion of nitric oxide to nitric acid involved extreme variations in temperature, and the acid formed tended to be very dilute. Two Americans, Robert Lovejoy and Charles Bradley, successfully produced nitric acid using the arc process in 1902, drawing on Niagara Falls hydroelectric power, but they were unable to acquire financing to move to commercial operations. B.A.S.F. also initiated research on the arc process, but the first Europeans to develop reasonably large-scale operations were the Norwegians Kristian Birkeland and Sam Eyde in 1903. The American rights to the Birkeland-Eyde process were later acquired by Du Pont.

Another individual interested in the arc process was Frank Washburn, an engineer who had been involved in nitrate mining in Chile. In 1907 Washburn, who had acquired a lease for a potential hydroelectric power site at Muscle Shoals, Alabama, went to Europe to investigate the Birkeland-Eyde operation. Discouraged by the huge amount of power required, Washburn learned of an alternative technique developed by two Germans, Adolf Frank and Nicodemus Caro, which produced calcium cyanamide; their method produced about 20 percent usable nitrogen by a process that combined nitrogen and calcium carbide. This "cyanamide process" also required large doses of electric power but less than was required in the arc process. Washburn bought the American rights to this process and persuaded the tobacco tycoon James B. Duke to bankroll the new venture. Washburn's company, American Cyanamid, leased power cheaply from Ontario Hydroelectric, a government-owned enterprise that was desperately seeking customers, and set up a plant at Niagara Falls in 1909; with its huge hydro power potential that area would soon become a favored location for the emerging American chemical industry.

A year later Washburn and Duke established another venture, Ammo-Phos Corporation, to produce ammonium phosphate fertilizer, incidently providing a big customer for Duke's phosphate mines in Florida. American Cyanamid was the largest commercial producer of nitrogen fertilizer in the United States on the eve of World War I and would figure prominently in the government's nitrogen program for that reason. But the cyanamide process shared the defects of the arc process: heavy power costs for relatively small returns in terms of usable nitrogen for fertilizer and military explosives.[11]

The search for a method of nitrogen fixation that would be both efficient and free of continuing heavy operating costs was ultimately concluded by two Germans, the chemist Fritz Haber and the B.A.S.F. engineer Carl Bosch. Haber, a professor at Karlsruhe technical institute in Baden, focused on a method of fixing nitrogen by combining it with hydrogen under controlled conditions of intense heat and extreme pressure to produce ammonia. The nitrogen content of ammonia is quite high compared to the amounts present in compounds formed by other processes, and the demand for electric power conversely lower. The major costs involved the design and construction of equipment that could withstand the extreme heat and pressure required in the process. In 1909, after completing experiments on a small scale, Haber approached B.A.S.F. for support in designing equipment to produce synthetic ammonia on a commercial scale. B.A.S.F.'s president Brunck assigned the task to his protégé, Bosch, who completed work on a small ammonia plant at Oppau, not far from B.A.S.F.'s big dye works at Ludwigshafen, in 1913.[12]

When war broke out a year later Germany soon faced a "nitrogen famine" similar to the "dye famine" on the Allied side. Her largest source of domestic nitric acid in byproduct ammonia from coke ovens was rapidly used up in the early campaigns, and the British blockade sealed off imports of Chilean nitrates that had supplied 50 percent of Germany peacetime demand. By the spring of 1915 the nitrogen needs of German farmers and the military had reached a crisis. Synthetic ammonia production at Oppau was expanded from 36,000 tons to 50,000 tons, producing about 12,000 tons of nitrogen, and the government initiated a crash program support for plants using the cyanamide process, but production continued to lag. In January, 1916, the German government in desperation agreed to finance the construction of a gigantic Haber-Bosch ammonia plant by B.A.S.F. at Leuna, near Leipzig, capable of producing 75,000 tons of nitrogen. By the time the Leuna plant was finished in the summer of 1918 it had almost three times that capacity and had cost over $15 million. B.A.S.F. emerged from the war as the largest chemical enterprise in the world, its premier technological position in synthetic ammonia augmented by a commercial alignment with the five other large German dye companies.[13]

On the Allied side the flow of nitrates from Chile was never interrupted, although a German naval squadron under Admiral von Spee posed a threat until it was destroyed off the Falkland Islands in December, 1914. German-owned nitrate mines in Chile were taken over and closed down by the British for the duration of the war. Most of the remaining nitrate fields were controlled by the British but there were also two recent American entrants: W. R. Grace and Company, an import house that had branched into mining in a small way in 1909; and Du Pont, which bought a nitrate mine, christened "Oficina Delaware," in 1910 to provide nitric acid for smokeless powder.

Du Pont had actually considered integrating into nitrate mining as early as 1903 to circumvent the prices imposed by an export cartel established in Chile in 1901. Inexperienced in international investment, faced with problems in acquiring local financial services in a country dominated by the Germans and the British, and alarmed by rumors of a possible Chilean nationalization of the fields, Du Pont backed away from an early commitment there. After the Panic of 1907, however, interest in vertical integration into nitrate mining revived as Du Pont discovered it had large inventories of nitrates and almost $2 million tied up in nitrate purchases at a time when cash reserves were essential. The nitrate exporter cartel was temporarily in disarray, but there was always the prospect of its resurrection. In this perspective a renewed effort seemed justified. Oficina Delaware actually supplied only 50 percent of Du Pont's nitrate needs but its existence and the threat of further expansion was used as a lever to control the prices offered by other suppliers.[14]

Although dictated largely by short-term financial considerations, the move into nitrate mining proved farsighted. Having secured its own source of supply, Du Pont was relatively immune to the pressure of rising prices created by increased demand by the Allies for nitric acid, and strikes in the Chilean fields which mine operators naturally blamed on German saboteurs. Aside from occasional problems in locating cargo space in ships coming from Valparaiso, a difficulty that was overcome once America entered the war and the War Shipping Board allocated priority space to Du Pont supplies, the flow of raw materials was steady. This circumstance enabled Du Pont to maintain a three months supply of nitrates throughout much of the war and thus keep her own military production in continuous operation. Du Pont's first substantial direct investment abroad, totaling assets of about $8 million by the end of the war, provided an essential element in ensuring the company's munitions production.[15]

The Chilean nitrate supply situation began to deteriorate in 1916. By the spring of 1917 there was a panic as prices on the spot market rose to 300 percent above normal levels, and an International Nitrate Executive was set up by the Allies to coordinate purchases and allocate movement of the material: Du Pont was assigned one third of the American share in this government-sanctioned purchasing pool. Meanwhile, proposals for government subsidization of synthetic nitrogen production surfaced, promoted particularly by congressmen from rural constituencies that were anxious to boost the fertilizer supply, since explosives absorbed most of the nitrate imports.

American Cyanamid was an early beneficiary of the renewed interest in atmospheric nitrogen. In 1914 the British government made an arrangement for Cyanamid's subsidiary, Ammo-Phos, to supply ammonium nitrate for Du Pont's smokeless powder production for the Allies. Meanwhile, Cyanamid's financial backer, James Duke, had been investigating the possibilities of expanding nitrate capacity by acquiring rights to the Norwegian Birkeland-Eyde process. He approached Coleman du Pont with a proposal for a joint venture in this field, using the hydro power he planned to develop on the Saguenay river in Canada. Pierre du Pont had been considering a similar undertaking, but not in conjunction with anyone else. In May, 1915, Du Pont set up a subsidiary, American Nitrogen Company, to develop the Birkeland-Eyde process in the United States.[16]

The problem of developing adequate electric power remained. The sites with the greatest potential in the United States were under the control of the federal government and development by private companies had been restricted since 1907. After some tentative efforts to acquire a Canadian power site, Du Pont decided to attempt to persuade the U.S. government not only to lease a power site to them but also to underwrite development costs. Pierre du Pont argued that, since his company "alone can consume in times of peace for commercial use the enormous quantity of nitric acid that the United States government might require for military use in times of war," it was logical that the electricity needed for nitric acid production be provided at public expense, although "the profit allowed to the manufacturer is to be decided by the government alone." To this end Du Pont arranged for Senator Willard Saulsbury of Delaware to present a bill along these lines despite Saulsbury's warning that "your own bill, coming from a concern of large wealth . . . in all human probability will be looked at with such suspicion and prejudice that it will not be able to obtain fair and disinterested considerations."[17]

Saulsbury's warning was prophetic. The bill, opposed by Democrats and Western congressmen who wanted new power sites developed in their own regions, was stymied in committee. Du Pont abandoned the project by 1917 although the company retained the Birkeland-Eyde process and remained interested in the technical and commercial potential of atmospheric nitrogen. In May, 1916, L. E. Edgar of Du Pont approached William Beckers, the dye maker, to determine if rumors that Beckers had access to the Haber process were accurate. This proved to be a futile errand, but Du Pont was attracted by the relatively low power requirements involved. Henceforth the Wilmington company focused efforts on procuring the ammonia synthesis process as the key to entering the nitrogen industry, a quest that would lead to increasingly complex and extensive contacts with foreign companies in the following decade.[18]

Meanwhile the United States government was moving, laboriously, toward a publicly supported program for the development of atmospheric nitrogen. In June, 1916, while the Saulsbury bill was being buried, Congress passed the National Defense Act which incorporated in section 124 a proposal drafted by Senator "Cotton Ed" Smith of South Carolina. This provision authorized an initial appropriation of $20 million to develop domestic nitrogen production: the facilities would provide nitrogen for nitric acid in wartime and fertilizer in peacetime. The section also required an investigation of various processes in use, a proposal that delayed the program for more than a year. The Nitrate Supply Committee of the War Industries Board, representing industrial interests, followed the recommendation of Eysten Berg of U.S. Army Ordnance who proposed an American effort to move into ammonia synthesis. The only American enterprise with any experience in this area was General Chemical Company, which had acquired American rights to a modified version of the Haber process in 1913; not surprisingly, William Nichols of General Chemical was a leading member of the Nitrate Supply Committee.

A committee established by the National Academy of Sciences under Arthur Noyes of M.I.T., however, maintained that successful implementation of a Haber process operation would take at least two years, a conclusion shared by Irenee du Pont who had examined the situation for his company at this time, and urged that the government provide hydroelectric power to support expansion of the well established cyanamide and arc processes.[19]

The judgment of the Nitrate Supply Committee initially prevailed and General Chemical was awarded a contract to build a nitrate plant at Sheffield, Alabama. Under the 1916 Act the plant would be operated by the U.S. government. This project, called Nitrate Plant Number 1, cost $12 million but never produced any nitrogen. As early as September, 1917, reports were circulating that the experimental plant would not be operating for at least another year and a half. By that time the war had ended and the plant was shut down. In December, 1917, the advocates of the cyanamide process, now vindicated, pushed through a new project, Nitrate Plant Number 2, to be built by American Cyanamid at the Muscle Shoals power site, with an annual capacity of 200,000 tons of cyanamide, equaling 40,000 tons of nitrogen. Work moved rapidly on this project, which was completed in less than a year at a cost of $68 million. Two more cyanamide plants were under construction nearby when the armistice halted all production. The U.S. government had spent $107 million and produced less than 100 tons of nitrogen.[20]

The government-owned nitrate works became a source of rancorous controversy over the next decade. In 1919 President Wilson appointed Arthur Glasgow, who had carried out a study of the European nitrogen industry, to be "Fixed Nitrogen Administrator" in charge of the plants. Glasgow first tried to interest private companies in buying the works, but even Washburn of American Cyanamid was reluctant to take over the expanded capacity in the face of an uncertain postwar market for fertilizer. Glasgow then proposed that the government operate the nitrate plants through a public corporation that would provide cheap hydro power as well as fertilizer to the region surrounding Muscle Shoals. The question of disposal of the nitrate plants thus became embroiled in the contentious issue of public ownership versus private power which submerged all other aspects of the situation. Eventually the Muscle Shoals hydroelectric works would become the centerpiece for the Tennessee Valley Authority.[21]

The American wartime experience with synthetic nitrogen thus proved to be something of a fiasco. Nevertheless, it played a part in promoting the development of the chemical industry in the postwar period. Du Pont began seeking access to the synthetic ammonia process. General Chemical benefitted from its mistakes in constructing Nitrate Plant Number 1, and a subsidiary established in 1919, Atmospheric Nitrogen Company, had a plant in operation at Hopewell, Virginia, by early 1921 at an estimated cost of $42 million. A year later Mathieson Alkali company had nitrogen works at Niagara Falls, based on a process developed by the U.S. Army's Fixed Nitrogen Research Laboratory. Du Pont began constructing its first ammonia plant at Belle, West Virginia, in 1924.[22]

The war period also stimulated amalgamation in the chemical industry, influenced in part by the large levels of investment required not only in developing the fixed nitrogen field, but also in dyestuffs, potash substitutes, acetone and benzol used as solvents in explosives, and related areas. The largest merger of the period was Allied Chemical and Dye Company, combining General Chemical with its large sulphuric acid capacity and its nitrogen process, Solvay Products Company and Semet-Solvay in soda ash, Barrett Company in coal tar products, and National Aniline in dyestuffs.

As in the National Aniline merger in 1916, the catalyst for Allied Chemical was Eugene Meyer, who continued to pursue the aim of creating an integrated chemical firm built around his dye company. At one point Meyer approached Du Pont but apparently more as a ploy in his negotiations with Nichols of General Chemical than as a serious proposition. Nichols had been involved in the National Aniline combination which had included a subsidiary, Benzol Products, but relations with Meyer's lieutenant, Orlando Weber, had never been good. Nevertheless, Meyer's persuasiveness and the prospect of a dominant position in the American chemical market overcame his misgivings. In September, 1920, the merger was announced, the new consolidated company holding assets of $283 million, compared to Du Pont's current asset value of $253 million.[23]

One other important merger of this period, though on a much smaller scale than Allied, was Union Carbide and Carbon Corporation, consolidated in 1917 with assets of about $150 million. This merger was simply the culmination of a steady process of interconnection among several Midwestern companies involved in the manufacture of carbon-related products. Union Carbide of Chicago had initially produced acetylene gas for illumination, then developed oxygen acetylene processes for welding and cutting metals, as well as manufacturing various kinds of metal alloys. The markets in these various fields expanded rapidly during the war, and led Union Carbide to closer ties with National Carbon Company of Cleveland, which produced carbon electrodes, and Prest-o-Lite of Indianapolis, manufacturer of acetone-filled packaging for Union Carbide's acetylene. The newly consolidated company continued its course of fairly broad-ranging industrial ventures on the boundaries of the chemical and metal industries. Its major chemical interests were in ethylene oxide, a base for solvents, and isopropyl alcohol, a base for acetone.

Subsequently Union Carbide moved into organic chemicals, specializing in those relating to natural gas and oil. In striking out on this course the Chicago company kept far afield from Du Pont and Allied, whose focus was on coal tar chemicals, and reflected the increasing range and complexity of the chemical industry.[24]

CHAPTER FIVE

Du Pont in Transition: The War Years

Du Pont's entry into the dyestuffs field in 1917, and the more tentative move into atmospheric nitrogen synthesis, were shaped in part by the short term opportunities in these markets generated by wartime demand. But these decisions were also part of a broader strategy of diversification into chemicals that Du Pont had been contemplating since 1908 when the prospect of a large nitrocellulose inventory loomed over the company as a result of government cutbacks in smokeless powder purchases. This particular danger proved to be ephemeral, and by 1915 Allied demands for Du Pont's explosives were straining the Wilmington company's nitrocellulose capacity. Nevertheless, Du Pont executives recognized that the wartime demand would be short-lived. They continued and expanded their diversification plans to ensure that their huge investment in military production could be absorbed into the production of nonmilitary chemicals. In that way, they hoped, they would not be faced with the problem of overcapacity that had endangered the company in the aftermath of previous wars. An awareness of history and a sophisticated understanding of the requirements of long-range business planning contributed to the success of this strategy.

Diversification: The Beginnings

From the time when they began reorganizing the internal operations of the company, the Du Pont cousins had recognized the need for long range planning. In 1903 a Development Department under Arthur Moxham (Pierre du Pont's association from the Lorain Steel years) was set up to monitor the activities of competitors, to ensure that future supplies of raw materials would be available for the manufacturing divisions, and to supervise research and development of new processes by the experimental laboratories. Pierre du Pont shared Moxham's interest in technical development as did Francis du Pont, who headed the Smokeless Powder Department, so that the work of this relatively small staff was assured of support from top management, particularly since Moxham and Pierre were both on the Executive Committee. During the first few years of its operation the Development Department was primarily involved in assessing the nitrate resource potential of Chile and related matters involving the acquisition of raw materials. After 1908, however, when Irenee du Pont took over the organization from Moxham, the Development Department increasingly became the focal point for planning for product diversification.[1]

The cutbacks in government orders of Du Pont smokeless powder in 1908, following in the wake of the antitrust suit, forced the company to give diversification more serious and sustained attention than it had earlier. The Executive Committee established a subcommittee, including Pierre and Amory Haskell, to investigate alternative markets for nitrocellulose so that the company could continue to run two of its three smokeless powder plants. The fields with the greatest potential were artificial leather, celluloid, and paints. The subcommittee and the Development Department decided that artificial leather, which was used for seat covers in automobiles, was the most promising.

In keeping with its tradition of expansion by acquisition Du Pont chose to buy a company already occupying the field, Fabrikoid Company of Newburgh, New York, rather than try to develop its own large scale production. In 1909 a small artificial leather operation was set up at Du Pont's inactive smokeless powder plant to provide the company with some production experience and which could also be used as a counter in negotiations with Fabrikoid. In the following year Du Pont acquired Fabrikoid, a company with assets of $780,000 and estimated net earnings of $250,000. In the negotiations Irenee du Pont demonstrated a flair for bargaining equal to Coleman's, persuading Fabrikoid's owners to accept payment of $1,195,000 in Du Pont preferred and common stock and a small amount of bonds in lieu of cash.[2]

In 1913 Du Pont extended its position in the artificial leather market by exchanging 16⅔ percent of Fabrikoid stock for an equal amount of stock in a British-owned firm, Pluviusin, a step taken partly to head off a potential competitor and partly to provide Du Pont with access to the foreign firm's technology. This was a tactic that Du Pont had traditionally followed in protecting its position in explosives, and set the pattern for future diversification initiatives involving foreign firms in such areas as cellophane, dyes, and synthetic fibers.[3] Du Pont subsequently bought several more firms in the field, expanding its assets in artificial leather to $7.3 million with net earnings of $800,000 by 1917 when the various companies were absorbed into the Du Pont Fabrikoid division.

The Executive Committee and the Development Department had concluded in its 1908–10 investigations that the time was not ripe to enter other fields. Du Pont's nitrocellulose capacity could be applied to the development of products based on pyroxylin, including celluloid plastics, lacquers, and artificial silk, but there were already too many small firms operating in what were still relatively small markets. Du Pont did approach a French firm, Chardonnet Company, which produced artificial silk from nitrocellulose, but the talks were inconclusive. In the meantime in 1911 a British company, Courtaulds, set up an American subsidiary, American Viscose Company, that produced artificial silk using a more efficient process.[4]

Nevertheless the Development Department continued research in these fields, and further government reductions in military purchases in 1911–12 renewed the Executive Committee's interest in diversification. In 1913 the Development Department urged a move into pyroxylin products, noting the rapid growth in demand for celluloid materials and the fact that Du Pont's large existing nitrocellulose capacity would ensure production costs well below those of smaller competitors. At the same time, however, Du Pont would probably have to enter this market on a vertically integrated basis, producing finished as well as semifinished products, since the manufacturers who bought these products were wary of "outsiders" acquiring information about their process and would demand "control and supervision of their product throughout its manufacture." Although this requirement suggested the need to acquire an existing company in the market, the Development Department recommended that Du Pont initiate its own operation to ensure integration of the pyroxylin venture within the Smokeless Powder Department and to employ idle plant and men. In 1914, however, when military demands for explosives absorbed the excess capacity Du Pont had only a small experimental pyroxylin plant, and in 1915 the Executive Committee decided to follow the pattern established in artificial leather, by acquiring the Arlington Company, the largest celluloid producer in the country. In this situation, as in the acquisition of Laflin and Rand, Du Pont benefitted from the circumstance that the aging major shareholders in the Arlington Company were increasingly uninterested in business affairs and divided over the choice of a successor. Du Pont paid $5.3 million for the Arlington Company out of its rapidly growing cash reserves.[5]

On the eve of World War I Du Pont had embarked on diversification on a major scale in artificial leather and was preparing to move into celluloid products. The rapid escalation of wartime demand for military explosives halted this program temporarily but did not fundamentally alter the direction in which the company was headed. If anything, the expanded smokeless powder capacity reinforced the need perceived by the Executive Committee for greater diversification after the war.

During this same period the Du Pont company experienced a bitter power struggle that divided family members, in some cases beyond hope of reconciliation. In the wake of this feud a new group of leaders emerged behind Pierre du Pont, including a number of younger, ambitious, and innovative executives. Their victory also produced a more tightly knit structure of control at the top of the corporation, a circumstance that would affect the postwar development of Du Pont, ensuring the rapid and thorough implementation of the diversification strategy and the reinvestment of a substantial proportion of the company's wartime profits in support of measures that transformed it into the leading chemical manufacturer in the United States.

The Struggle for Control

Although the controversy that led to the "Du Pont civil war" did not come to a head until 1915, the seeds of discord had been germinating for almost a decade. The three cousins were at the center of what appeared on one level to be a contest over the succession of control of the Du Pont company, but also involved long-festering personal animosities among other family members and dissatisfaction among younger company executives—many of whom were also linked to the family by blood or marriage—over the management of the company.

Of the three cousins who had reorganized the Du Pont company in 1903, Alfred I. du Pont had been the least involved in the business beyond managing production in the Black Powder Department. In 1906 he resigned this position and also temporarily his place on the Executive Committee. During this time Alfred divorced his first wife and remarried, arousing the ire of a number of Du Pont kin who prevailed on Coleman to persuade Alfred to absent himself from Wilmington. Although Alfred returned to the Executive Committee in 1909 the tensions generated by family bickering continued. Meanwhile, Coleman, plagued by illness and lured away from Wilmington by real estate interests in New York and the prospects of a political career, gradually withdrew from company affairs. In 1911, as the specter of the antitrust judgment loomed over the company, Coleman decided to reorganize top management, to reduce the Executive Committee's routine workload and give it an enlarged planning role, and to provide the younger executives greater authority in running the operating departments. Pierre went along with this scheme, partly perhaps because it provided greater opportunities for his younger brothers, Irenee and Lammot; but Alfred objected vehemently, particularly to the proposal to install Hamilton Barksdale as general manager of the operating departments, including Black Powder. Alfred's criticism of the plan was borne out as Barksdale, though a capable subordinate, proved unwilling to delegate authority to the managers of the operating units, defeating a main purpose of the reorganization effort. In 1914 the company was reorganized again, eliminating the position of general manager and the plethora of committees that had been established under the Executive Committee, restoring direct links between the Executive Committee and the operating departments.[6]

The failure of Coleman's reorganization effort and his near-disastrous intervention in the negotiations over the antitrust settlement in 1911–12 strained his relations with Pierre at the same time that Alfred's personal affairs created tensions throughout the family. A further dispute arose when William du Pont, another family member with large shareholdings in the company, sought a more active role in directing the firm. Coleman supported him but Pierre objected on the grounds that William had little business experience or qualifications. Pierre eventually compromised on this issue but brooded over the "insecure grasp that he and his ideas had on the company." The prospect of the firm falling under the influence of Alfred or William who "made decisions . . . on the basis of snap judgments and personal whims" he found unsettling. At the same time he wanted to preserve the tradition of family control of the company.[7]

Pierre's opportunity came in 1914. Coleman was now definitely committed to a major real estate investment in the construction of the Equitable Life Insurance Building in New York and needed $8 million in capital. He had also become ill and had to go to the Mayo clinic twice during the year for operations. Coleman shared Pierre's concern that the younger executives should be given a greater share in the company as an incentive to hold them. Moxham was already planning to leave to take over a new venture, Aetna Explosives Company, that he was organizing with several younger Du Pont cousins. Many of the new members of the Executive Committee did not hold substantial stock in the company. For these reasons in December, 1914, Coleman wrote to Pierre proposing to sell 20,000 shares of Du Pont common stock, about 49 percent of his total equity in the company, for $160 per share. The stock could then be distributed to members of the Executive Committee and other Du Pont managers earning over $6,000 a year.

Alfred objected to the proposal. At this point Allied munitions orders had not begun to have an impact on company sales and Du Pont common stock on the New York market hovered around $120 per share in the fall of 1914.[8]

Alfred, supported by William du Pont, prevailed on the company's finance committee to reject Coleman's offer at any price over $125 per share.

Matters remained at a stalemate for several weeks after the Finance Committee meeting. Then Pierre heard from a Nobel representative who was visiting the United States that there were rumors in London that Kuhn, Loeb and Company, a New York investment house suspected of pro-German sympathies, might acquire a substantial block of Du Pont shares. By this time Allied orders were flowing in to Du Pont, incidentally boosting the value of Du Pont common stock, and Pierre was alarmed that the rumors, which could only involve Coleman's shares, might jeopardize relations with the British government. Initially Pierre, with Alfred's approval, proposed that all the major family shareholders should pool their stock so that there would be assurances that none of it would be sold or used as collateral for loans for the duration of the war. Coleman rejected this idea, but countered with an even more startling proposal. He was now prepared to sell all of his Du Pont stock for approximately $200 per share.

Rather than taking this proposal back to the Finance Committee, Pierre assembled a small group of Du Pont executives, including his brothers Irenee and Lammot, another cousin, Felix du Pont, and two close associates, R. R. M. "Ruly" Carpenter and John J. Raskob. They agreed to set up a syndicate to buy Coleman's shares by borrowing $8.5 million from a J. P. Morgan affiliate, Bankers Trust of New York and pledging their own shares of Du Pont common stock as collateral. The group established a corporation, Du Pont Securities (later renamed Christiana Securities Company), to make the purchase. The entire transaction was completed within five days from the date Pierre received Coleman's letter, and neither the Du Pont company nor the other family members learned of it until the news was released a week later, on February 28, 1915. Pierre and the syndicate now controlled the largest block of shares in the explosives company.[9]

Alfred was outraged when he heard the news. Both Alfred and William du Pont demanded that Coleman's shares be turned over to the Du Pont company, arguing that Pierre and his syndicate could only have acquired them by pledging company stock as security for a loan, which was true. Pierre, however, initially rejected the demand, arguing that the Finance Committee had already declined to buy Coleman's shares at the lower price Coleman wanted in December. Even historians sympathetic to Pierre have admitted that "the reason Pierre moved so fast and secretly . . . was that he wanted to obtain control of the family company," as he "had come to feel that he could not work with Alfred and William."[10] In extenuation of his actions, however, they note that on the same date Pierre received Coleman's second offer, Alfred had written to Coleman insisting that $125 per share was a fair price for the stock. Subsequently Pierre arranged for the distribution of 1,250 shares in Christiana Securities to powder company executives, and offered shares in the venture to other family members on the same terms as the original members of the syndicate. Nevertheless, this distribution of shares did not dilute Pierre's control, and in early March, 1915, he was elected by the powder company board to succeed Coleman as president.

Pierre also agreed to allow the powder company board to entertain a motion by Alfred to purchase Coleman's stock, anticipating correctly that they would reject it. Alfred and his allies were temporarily stymied, but in December, 1915, they counterattacked with a lawsuit initiated by the minority shareholders. They alleged that Pierre had withheld information from the Finance Committee in its December, 1914, meeting concerning the new Allied contracts that would shortly boost the powder company's common stock from $120 per share to $775 per share by September, 1915, when the stock was split at two for one. The stock rose again to $430 per share by October, 1915.[11]

While Du Pont's wartime earnings soared the case wound its way through the U.S. District Court in Delaware. The internal struggles of the increasingly wealthy and powerful company were recounted with relish in the press, to the distress of this traditionally publicity-shy family. At the annual stockholders meeting in March, 1916, Alfred, who had already been removed from active management, William, and Francis du Pont, who had taken Alfred's side in this dispute, were all purged from the board of directors.

Finally in April, 1917, the court reached its decision. Despite Coleman's testimony that his offer in February, 1915, had been to Pierre, not to the Du Pont company, the judge upheld Alfred's position and ordered that a special shareholders meeting be called to reconsider the motion to purchase Coleman's stock. The decision also enjoined Pierre, whose conduct in the transaction was characterized as bordering on fraud, from voting the stock acquired by Christiana Securities. Despite this setback, and the bad publicity, Pierre and his allies in the company and family mobilized enough votes to reject the proposal in October, 1917, and the case was dismissed. When Alfred appealed the dismissal, the U.S. Circuit Court in March, 1919, rejected his plea, and also criticized the lower court for impugning the integrity of Pierre and his associates. After five years of bitter acrimony, Pierre emerged victorious, his control of the company reaffirmed and reputation more or less restored.[12]

The "Du Pont civil war" had been extremely divisive and left enduring scars on family relations. Alfred moved to Florida and had nothing further to do with his erstwhile kinsmen beyond supporting candidates opposing Coleman when the former company head embarked on a political career in the

1920s. Francis du Pont was later reconciled with Pierre, but others who had supported Alfred remained embittered, and the old battles were still being re-fought in various historical accounts a generation later.

Judged in terms of the company's fortune and future development, Pierre's triumph was probably a foregone conclusion, given the rising value of Du Pont stock which soothed family sensibilities, and was beneficial in the long run. The coup of February, 1915, ensured that the company would be run in the immediate future by a single-minded and well coordinated group of managers whose individual fortunes were clearly tied to the success of the firm. A company divided at the top might have moved with less alacrity to take advantage of the opportunities present in the war years. Executive loyalty was augmented by the expansion of the stock bonus arrangements carried out by Pierre in 1915. Alfred had endorsed the concept of wider distribution of shares in principle but balked at measures to put the idea into practice. Finally, the concentration of ownership in the hands of those involved in active management of the company ensured that a substantial proportion of the earnings from war-time munitions production would be reinvested in long-term diversification of Du Pont into chemicals rather than distributed to the shareholders. Although Pierre had been one of the major figures involved in the earlier reorganization of the power company, his assumption of control in 1915 marks a turning point in the direction of Du Pont toward diversification and the emergence of a new generation of leaders in the firm.

World War I: Diversification Renewed

When war broke out in Europe in August, 1914, Du Pont and much of American industry was in a recession. The company's gross earnings had declined by about $9 million from its high point in 1912, a situation aggravated by the division of assets under the antitrust decree. Allied military orders began coming in within two months but Du Pont reacted cautiously. Most of the orders were for explosives that Du Pont did not manufacture in large quantities, such as picric acid and T.N.T., and any major expansion in production would require substantial new capital investment. Many Americans as well as Europeans shared the belief that war would end fairly quickly, and Pierre du Pont feared being stranded with excess capacity. Consequently, Du Pont insisted that the price it received for explosives bought by the Allies should be high enough to cover the costs of expansion; and that 50 percent of the purchase price should be paid up front, when the contract was signed, and an additional 30 percent when the powder was midway through processing. The Allies would absorb the full cost of new investment to enlarge capacity as the final payment would be made when the powder was ready for shipment. Since Du Pont was the largest producer of military explosives in the United States and had safe as well as direct control over raw materials such as nitrates, the buyers had little choice but to go along. In October, 1914, the French government entered a contract drawn on these lines for 9 million pounds of powder.

Others followed suit. Within one year Du Pont's gross receipts had quadrupled to more than $130 million which covered the entire amortization cost for plant expansion and left a staggering net of $57 million, larger than the company's gross earnings in its best prewar year. By 1916 gross receipts had again more than doubled and the company's net earnings in that year equaled $82 million, some $20 million more than Du Pont's accumulated net earnings for the entire period from 1904 to 1914. A congressional investigating committee in the 1930s would charge Du Pont with exploiting its position to extract excessive profits from the Europeans, and later the United States government for military production. Du Pont spokesmen maintained that the high prices assessed against the Allies were necessary to finance expansion and could point to the contrary experiences of companies such as Moxham's ill-fated Aetna Explosives and the Winchester Arms Company, both of which expanded capacity with borrowed capital and ran into problems later when Allied contracts were canceled or altered. Furthermore, Du Pont argued that after mid-1916 powder prices leveled off as the company reached full production just in time to begin filling orders for American military needs. Because the company had enlarged capacity at no extra cost to itself, and had inventories of raw materials, powder prices for the U.S. government remained stable through 1917–18 and well below the costs of other commodities, including the market prices of raw materials used in military explosives.[13]

More significant in terms of Du Pont's future development, the wartime expansion, executed on a self-financing basis, provided the company with a huge reserve of $89 million in retained earnings for investment in diversification; and the wartime production left the company with a staff of managers, chemists, and engineers experienced in the uses of nitrocellulose materials, and new plant and equipment available for exploitation. In 1915 the Development Department was assigned the task of identifying products whose manufacture could be readily undertaken in the new war plants as military demand slackened. Within two years the Development Department was given an expanded responsibility, to prescribe a diversification strategy that would ensure the most effective use of all the company's resources and set a course for Du Pont's postwar development.

Initially the Development Department concentrated on finding new uses for the smokeless powder plants in New Jersey and a huge Hopewell, Virginia, plant that had been built to meet the Allied war demand. The department's head, R. R. M. Carpenter, an M.I.T.-educated associate of Lammot du Pont who had taken over the position from Irenee du Pont in 1911, submitted reports in January and May, 1916, and a more elaborate statement of the general plan at the end of that year, on the eve of America's entry into the war, which would provide for another infusion of capital and further pressure for expanded capacity.

Some of the war plants, particularly Hopewell, were considered too large to be used effectively after the war. Each of the other plants was assessed in terms of its capacity to adapt to new product lines. Finally, the department proposed developing "a system of related industries" which would employ "a material proportion of the excess plants within . . . three to five years," and would ultimately absorb as well the skilled technicians, equipment, and capital that Du Pont had accumulated during the war. These "related industries" included vegetable oils and their byproducts, such as soap; paints and varnishes; dyestuffs and related organic chemicals; water soluble chemicals; and synthetic textiles and similar products based on nitrocellulose fibers.[14]

To some extent the choice of fields for diversification was fortuitous. The decision to enter dyestuffs, for example, had been made by enthusiasts on the executive committee such as Irenee du Pont despite objections from the Development Department. At the same time, however, there was an underlying technical as well as commercial logic to these proposals. Each of the products considered was manufactured from materials that were, or could readily be, produced in existing Du Pont plants. Many of the products were interrelated: water soluble chemicals such as tartaric acid and oxalic acid were used in dyes; linseed oil was used in paints; other forms of vegetable oils were ingredients in soaps and detergents. Under this plan Du Pont would emerge not only as a diversified chemical company but also would retain many of the benefits of vertical integration that had helped ensure its dominance in the explosives industry.

In keeping with the company's tradition of financial prudence, Du Pont imposed an additional test of commercial feasibility on each new diversification proposal. Du Pont followed a standard requirement that any major new ventures should be able to guarantee a return on investment of at least 12 ½ percent within a reasonable length of time. Performance was carefully monitored under the procedures initiated by Pierre du Pont in his term as company treasurer. This is not to say that the company never made a poor investment decision. The most egregious example is dyestuffs, particularly the ill-fated venture into the export market, discussed below. The paint and varnish business also ran into rough sledding in the postwar period. Nevertheless, the rigorous standards applied by Du Pont ensured that their enterprise would avoid some of the pitfalls that awaited other companies entering high technology fields, in particular a preoccupation with exploiting the sales potential of new markets or with technological innovation regardless of cost.[15]

The Development Department also recommended a continuation of the practice of expansion by acquisition: as in the case of celluloid and fabrikoid, immediate access to an existing firm's know-how was considered preferable to trial and error experimentation. Du Pont's great advantage now lay in its size and capital resources, and the consequent economies of scale that it could achieve. Early in 1917 Du Pont bought Harrison Brothers of Philadelphia, an $8.5 million firm specializing in paints and pigments, but also involved in the manufacture of acids and related heavy chemicals. Later that year Du Pont extended its move into paints, purchasing 80 percent of the Flint Varnish and Color Works, a company that specialized in automobile finishes. By 1920 Du Pont had invested $12 million in paints and varnishes.[16]

When no going concern was available, the Development Department initiated its own research program. In 1917 Carpenter assigned a study of the vegetable oil field to Dr. Fin Sparre, a Norwegian chemist who had come to Du Pont originally to work on the Birkeland-Eyde nitrogen process, but soon exhibited considerably broader abilities and was later to take a major role in the complex technical negotiations with foreign firms.[17] The department also continued to investigate the potentialities of artificial fibers despite the earlier rebuff of Chardonnet Company. Meanwhile, expansion in existing lines was maintained. As noted earlier, in 1915 the Arlington Company was acquired, giving Du Pont a foothold in the wide-ranging plastics field. Arlington's existing management was retained but the company was placed under the stewardship of Lammot du Pont, Pierre's youngest brother and of the three who now dominated Du Pont probably the best trained technically. Lammot was also given charge of Harrison Brothers when it entered the Du Pont domain.

America's intervention in the European war in the spring of 1917 temporarily slowed the diversification drive as the company's resources were directed to meeting the new surge of military demand. The top management was further distracted by a prolonged controversy with the U.S. government over the prices charged for powder and the estimated cost of a proposed giant powder plant at Old Hickory, Tennessee.

Although the Wilson administration had begun mending fences with the business community as the prospects of war loomed, relations between the government and the Wilmington powder company were strained in 1916 by the controversy over Du Pont's proposed synthetic ammonia venture, and a tax surcharge imposed on "munitions profits" that Pierre du Pont believed was

The "Old Hickory" plant in 1918. Section of a panoramic photograph. *Courtesy of*

specifically directed at his firm. Nevertheless, when war came Du Pont anticipated that, as the largest explosives producer in the country, it would be involved in any plans for expansion of smokeless powder capacity undertaken by the government. Army Ordnance officials did in fact rely on Du Pont supplies in its initial wartime purchases and consulted with Du Pont when the government decided to expand capacity in the fall of 1917. At this point, however, the issue became enmeshed in bureaucratic and partisan politics in Washington. Secretary of War Newton D. Baker, acting on the advice of Robert Brookings of the civilian War Industries Board, rejected the contract drafted by Army Ordnance with Du Pont for the construction of the Old Hickory plant on the grounds that the cost estimates were too high and would provide overly generous profits to Du Pont. The dispute simmered for several months until the W.I.B. concluded that no other contractor could complete Old Hickory in the time required; and Du Pont agreed to reduce its commission for the job from $1 million to $500,000, and to accept a lower per-pound payment for the powder produced there. In the end, however, Baker had to renegotiate the contract to expand Old Hickory's capacity to meet the unprecedented military demand for explosives in the final campaigns of 1918. Du Pont made a gross profit of $2.7 million from Old Hickory, although the net after excess profits taxes were imposed came to $632,000.[18]

Du Pont's executives were never totally reconciled to the resolution of the Old Hickory controversy. The company concluded later that it had in fact made a patriotic sacrifice in this venture, as the return on its $129.5 million investment after taxes and stock bonuses paid to employees was less than 1 percent. At the same time Du Pont had little cause for complaint about the general state of earnings from the war years. According to the company's accounts, its gross receipts in 1917–18 came to $599 million, with net earnings of $92.4 million, and its accumulated surplus after bond interest and dividends was $68 million. Total assets had increased to $308 million, more than three times as large as in 1914. Part of this total represented installations such as Hopewell that would have little value after the war; but since these properties had been largely amortized Du Pont did not have a serious debt load. The bulk of the retained earnings was available for development of the new chemical fields entered in 1914–16, and for further ventures in diversification.[19]

The most important new venture undertaken by Du Pont in 1917–18, and in the long run one of the most significant to the company's future development, was the investment in General Motors. At the time the initial investment decision was made the potential value of the automotive firm was not as apparent as it seems in retrospect. Established in 1908 by William C. Durant through the merger of a number of small auto manufacturers, General Motors had experienced a checkered career. Durant had overextended himself financially in 1910, and the company was taken over by a banking syndicate under James J. Storrow of Lee, Higginson and Company of Boston which had pursued an extremely cautious course, declining to follow the lead of Henry Ford in exploiting the burgeoning mass market for automobiles. Meanwhile Durant, ousted from direct management, schemed to recover control of G.M. One of the men approached by Durant as a possible ally in his takeover strategy was John J. Raskob, Pierre du Pont's closest associate and his successor as treasurer of Du Pont in 1913. Raskob persuaded Pierre and other members of the Christiana group to invest in G.M. As the auto company's stock began to rise in 1915 the Du Pont executives increased their purchases. At a board of directors meeting in September of that year, Pierre du Pont was offered the chairmanship of G.M. and invited to act as "mediator" between the Storrow and Durant factions. By the summer of 1916 Durant had recovered control of G.M., partly through the financial backing of the Du Ponts and partly through a complicated stratagem in which he offered shares in his new and highly profitable company, Chevrolet, in exchange for G.M. stock. The bankers departed and Durant asked Pierre du Pont to remain as chairman while he resumed the presidency of G.M.[20]

Preoccupied with running his own company, Pierre had little time for G.M. initially, an arrangement that suited Durant. Meanwhile, Raskob urged his fellow members of the Du Pont executive committee to consider a direct corporate investment in the auto company. Du Pont's expanding capacity in artificial leather, and the new move into paints in 1917 enhanced the appeal of that investment considerably, and Raskob added the argument that G.M.'s postwar expansion plans would help absorb some of Du Pont's expected surplus of engineers and construction crews. There was opposition from some quarters: Lammot du Pont and R. R. M. Carpenter resisted proposals that would divert funds from the chemical ventures then being advanced by the Development Department; and Edmund Buckner, who headed the Explosives Department, foresaw an imminent demand for expansion in his area. Raskob overrode these objections, noting the uncertainties surrounding the Old Hickory contract and emphasizing the market potential of G.M. for Du Pont's chemical products. After much debate, Du Pont's executive and finance committees agreed to authorize an initial purchase of $25 million of G.M. stock in December, 1917; by the end of 1918 the investment had increased to $42.5 million, more than 60 percent of the retained earnings reserve.[21] At that time Du Pont held 3,882,906 common shares of G.M., controlling approximately 24 percent of the company's voting stock.

Du Pont was not the only large institutional investor in G.M. The Lee, Higginson group retained a large block, as did J. P. Morgan's banking syndicate, which had developed close ties with Du Pont in its financing of Allied munitions purchases. Du Pont, however, took an increasing interest in the management of the auto company, principally through Raskob, who became chairman of the new G.M. finance committee, modeled on Du Pont's, that was imposed on Durant in 1918. Durant and Raskob planned a major expansion of G.M.'s capacity after the war, but Du Pont was now reluctant to divert any further reserve funds to the auto company as demands for further support of chemical ventures were more pressing. Instead Pierre du Pont and Raskob approached Harry McGowan of Nobel, proposing that the British firm, which was embarking on its own program of diversification, invest in G.M., which was at that time laying the groundwork for a major export campaign.

At this point McGowan was not prepared to take the plunge, and Du Pont picked up a small amount of G.M. stock, increasing its total investment to $48.8 million. In the following spring, however, Nobel did follow Du Pont's lead and agreed to purchase 1.8 million shares through its subsidiary, Explosives Trades Ltd., and the joint Du Pont–Nobel Canadian affiliate, Canadian Explosives Ltd. (C.X.L.). The economic slump of 1920 hit Nobel, Du Pont, and G.M. simultaneously, and forced McGowan to back off from his initial commitment of $36 million, but Du Pont negotiated a short-term arrangement with the British firm under which C.X.L. would absorb an additional $8 million beyond its original $6 million share of the issue, with the understanding that Nobel would eventually take over this commitment. Despite the initially troubled conditions that surrounded these transactions, the long term benefits of the G.M. investment for both Du Pont and Nobel reinforced the increasingly close ties between the two chemical giants.[22]

The brief but severe postwar depression also created serious internal problems for Du Pont, as related below, and a crisis in the relationship between Du Pont and G.M. Durant's approach to business affairs, emphasizing sales volume over return on investment, had always run counter to the cautious Du Pont philosophy; and Durant was an exceptionally poor manager, allowing the various auto companies absorbed into G.M. to operate virtually free of supervision, allocating new investment funds with little assessment of their profit potential. During 1918–19 Pierre du Pont and Raskob had attempted, with little success, to prevail on Durant to reorganize G.M.'s affairs on a more orderly basis. The 1920 slump caught G.M. in the midst of its costly expansion program with a large inventory of cars, and jeopardized Du Pont's substantial investment. When Durant admitted that he had been borrowing heavily, using his G.M. stock as collateral, in order to buy up G.M. shares above market prices in a vain effort to stem its collapse, Pierre du Pont moved rapidly, taking over Durant's G.M. stock and ousting him as president. Pierre had retired from active management of the Du Pont company in 1919, but now took over the reins at G.M. to reorganize the auto firm and save it from disaster. With the assistance of some borrowed Du Pont men, such as Donaldson Brown, and junior G.M. executives, in particular Alfred Sloan, Pierre carried through his program, providing G.M. with a solid base for growth in the 1920s.[23]

Du Pont emerged from this crisis with more substantial control of G.M. The combined holdings of the Du Pont company and the Christiana Securities group was estimated at 38 percent of the voting stock of G.M. in 1923. Subsequently Du Pont gradually sold off a large portion of the stock acquired in 1920–21, so that by 1938 its holdings in G.M. were back to about 24 percent of the outstanding shares. Nevertheless, the block was large enough to arouse the interest of the antitrust division of the Justice Department in the 1920s and again after World War II. Antitrust lawyers were to lay special emphasis on the "captive market" that G.M. provided for Du Pont Fabrikoid, paints, and plastics. Du Pont spokesmen argued that their company never adopted a policy of imposing their products on G.M., but the potential value of G.M. purchases was clearly a persuasive element in the debate within Du Pont over investment in the auto firm in 1918, and the close links between the two companies encouraged the practice. In 1947 G.M. was buying 70 percent of its paints from Du Pont, and G.M. sales represented more than 80 percent of the business of Du Pont's Fabrics and Finishes Department.[24]

The major benefit that the G.M. investment provided Du Pont, however, was in return on investment, both from dividends and the escalating value of G.M. stock. The auto company under its new management rebounded rapidly from the 1921 depression and had displaced Ford as the leader in sales and earnings in the field by 1927. By the end of the decade G.M. dividends provided more than 50 percent of Du Pont's net income, in effect cushioning the chemical company against losses in such fields as dyestuffs and relatively slow growth in other lines. Du Pont's earnings from G.M. also contributed to the Wilmington firm's cash reserves, enabling it to continue its expansion and diversification after the war surplus reserve was exhausted, without substantially increasing its debt load through bank loans or stock issues that would weaken the equity position of the Du Pont family and the Christiana group. According to estimates made at the time of the Du Pont–G.M. antitrust case, G.M. contributed a total of $233 million, or one third of Du Pont's total net income, in the period 1920–52, during which time the book value of Du Pont's equity in G.M. increased from $48.7 million to $451 million.[25]

CHAPTER SIX

Du Pont in Transition: The Postwar Era

When Pierre du Pont stepped down as president in May, 1919, he left as a legacy a company that had been twice transformed: first in 1903–5 from a member of an explosives cartel into an integrated organization dominating the industry in the United States; and then in 1914–18 from an explosives firm into a diversified chemical company with assets of more than $230 million and earnings exceeding $17 million. This growth and diversification had been accomplished without any serious dilution of family control over the firm; if anything the degree of ownership control had been concentrated in the hands of one branch of the family—Pierre and his brothers—and the active management through Christiana Securities equity in the Du Pont corporation. In this feature, Du Pont differed markedly from other emerging large chemical firms, such as Allied Chemical and Union Carbide, which had achieved growth through merger, a relatively wide distribution of shares and a larger burden of long-term debt.

Du Pont also differed from other American chemical companies in its range of products, including industrial supplies such as acids and dyes, and consumer goods such as paints and celluloid plastics. During the 1920s the large chemical enterprises in Britain and Germany were to develop along similarly diversified lines, and smaller American companies such as Hercules Powder, American Cyanamid, and Union Carbide would branch into chemical consumer products. At the beginning of the decade, however, Du Pont occupied a rather unique position in the industry: in terms of assets and sales, it was second only to Allied Chemical; in many of its diverse product lines it was unquestionably the largest firm in the United States market.[1]

Pierre du Pont continued as chairman of the board of Du Pont, a position he did not relinquish until 1940. He also remained as chairman of the board of

Christiana Securities through the 1930s, so his indirect influence on Du Pont endured long past his departure as an active chief executive. In 1921, however, the crises at General Motors led Pierre to take on the presidency of the beleaguered auto firm in order to direct a reorganization effort, and his attention was increasingly diverted away from Du Pont affairs. Leadership thus passed in practice as well as in form to Pierre's younger brother, Irenee, who succeeded him as president in 1919 and continued to exercise considerable influence on the company even after he stepped up to vice-chairman of the board in 1926.

Personally more affable and extroverted than his older brother, Irenee brought to the position a different set of talents and in certain respects a different outlook from Pierre. Pierre was quiet, meticulous, and emphasized financial routines. Irenee bore a closer resemblance to cousin Coleman. He "liked night life, the ladies, good stories," and the excitement of negotiating deals. He was to find in Harry McGowan of Nobel a kindred spirit, a circumstance that contributed to the increasingly close connections of the American and British chemical giants in the 1920s. Irenee was also fascinated by technology. While he lacked the rigorous and systematic approach to scientific matters that was characteristic of his brother and successor, Lammot du Pont, Irenee had been educated at M.I.T., and had headed the Development Department when it initiated the diversification program in 1908–12. Later, Irenee would cause the family some embarrassing moments when he publicly advocated the use of mind-stimulating drugs to increase workers' productivity. During the Nye committee hearings into the arms trade in the 1930s Irenee's obvious contempt for the Senate investigators contributed to Du Pont's public image as arrogant "merchants of death." At the same time, Irenee was as farsighted as Pierre in charting corporate long-term development during his term of office and demonstrated a flair for public relations as lobbyist for the chemical tariffs. As president of Du Pont, Irenee contributed to the renewed emphasis on technological innovation and a distinctly novel awareness of the international aspects of the chemical industry.[2]

Irenee du Pont's succession to the presidency also marked the emergence of a new group of executives at the head of the corporation. Among the most important figures in setting the direction of Du Pont in the next decade were R. R. M. Carpenter and his younger brother Walter, both of whom had been involved in the Development Department during the war years, along with Lammot du Pont, Pierre's youngest brother; William Coyne and Frederick W. Pickard, both of whom were from sales backgrounds, and were oriented toward export markets; Harry Fletcher Brown, a chemist who had directed the expansion of Du Pont's smokeless powder and nitrocellulose operation from

1911; and Frank Tallman and J.B.D. Edge who had been involved in Du Pont's increasingly complex international purchasing operations during the war. A number of these managers (including the Carpenters, Brown, Tallman, and Coyne) had significant personal investments in the company through shares in Christiana Securities. Many of them had been involved in the diversification program and reflected Du Pont's new orientation as a chemical enterprise. They also had a much wider experience in international markets than the men they were replacing on the Du Pont executive committee, and supported the more outward-looking approach that would be a distinguishing feature of Irenee du Pont's term as president.[3]

Crisis and Reorganization, 1919–21

The problem that confronted Irenee du Pont most immediately upon his assumption of the presidency involved domestic sales rather than foreign markets. Despite the cushioning effect of war profits, Du Pont faced some serious financial difficulties that were compounded by administrative deficiencies. Although the postwar fluctuations of the U.S. economy, which culminated with the 1920–21 depression, contributed to this situation, the problems were primarily internal, the result of the headlong move into diversification in the war years.

Before the war, Du Pont sold its products to two kinds of customers: explosives to miners and industrial contractors, and smokeless powder to the U.S. military. Purchases were usually made in large quantities and the emphasis was placed on the uniform quality of the product. Du Pont was therefore concerned primarily with maintaining the flow of materials and standards of production; marketing posed few problems. When the company was reorganized in 1903–4, the main objectives had been to centralize production and sales to promote coordination that would ensure the continuous controlled movement of a large volume of standardized explosives products. Du Pont was organized on a "functional" basis with departments established to manage each phase of this process: purchasing the materials, property and real estate, production, sales, and research and development. Given the relative homogeneity of the explosives market, sales could be coordinated with manufacturing on a centralized basis from the Du Pont headquarters in Wilmington.[4]

As Du Pont began to move into new fields, marketing problems multiplied. In dyestuffs, for example, most buyers were small textile firms that wanted small amounts of specialized materials rather than a large volume of standardized materials. The difficulties in filling this wide variety of orders were compounded by technical problems encountered in producing dyes of uniform quality, and contributed to high costs and unexpectedly slow development of this field.

Paints and plastics presented even more complicated marketing difficulties. Both fields involved bulk sales to other manufacturers and retail sales, called "merchandising," in which Du Pont had relatively little experience. The Arlington Co. in celluloid products and Harrison Brothers in paints were largely oriented toward this second market. Du Pont had assumed that extending its successful methods of production control would reduce costs in these areas so that earnings would increase even without the expected substantial increase in sales volume. As early as 1918, however, it had become clear that these expectations were not being fulfilled, particularly in the paint business. Sales were increasing but the return on investment declined. As the general manager of the department admitted, "the more paint and varnish we sold, the more money we lost." Du Pont hoped to stem the decline by restricting further acquisitions in the field to companies such as Flint Varnish that were primarily industrial suppliers but, paradoxically, "this industrial business was immediately allowed to slip from our hands and our sales efforts concentrated on trade sales."[5]

As earnings continued to plummet in 1919, Pickard, the vice-president in charge of sales, concluded that the basic problem for Du Pont was its unfamiliarity with retail merchandising. Market demand for the various products was growing, but due to poor coordination of sales efforts, Du Pont was losing business to smaller competitors more familiar with each individual field. The company faced the alternatives of leaving these fields completely or embarking on expensive programs to establish national advertising and distribution methods throughout the country.

The Du Pont Executive Committee responded in its customary methodical fashion to the problem, establishing a study group that consulted with a variety of firms with more experience in retail markets. This investigation led the members of this group, mostly junior executives with experience in operations rather than general policy, to conclude that while there were similarities in production requirements, each field faced distinct marketing problems. What was needed was a reorganization of the company with separate divisions for paints, chemicals, and pyralin (plastics), each division exercising a good deal of autonomy in devising sales and distribution strategies. In short they would "make product rather than function the basis of the organization" of Du Pont, and provide a substantial degree of management decentralization.

These proposals encountered resistance from senior members of the executive committee, including Irenee du Pont, who saw little merit in decentralization and the apparent abandonment of the system of functional coordination.

As a compromise the advocates of decentralization recommended establishment of "steering committees" that would devise special strategies for the fields where marketing problems were most severe, beginning with paint and varnishes. During 1920 and early 1921 this approach was extended to most of the major product lines of the company. At this point, however, the postwar depression began to take effect. Every field except explosives showed losses in the first six months of 1921: dyestuffs showed a deficit of $1 million and even traditionally profitable lines like Fabrikoid suffered almost as badly. The net loss came to $2.4 million and the whole strategy of diversification appeared in jeopardy.

Du Pont's troubled top management, including Pierre du Pont and Raskob, met in August, 1921, to review the situation. To them the company's long term future seemed precarious. Although the company's debt position was manageable and losses could be absorbed, the new investments on which postwar growth depended were all in trouble. The cushion of retained earning from war production was seriously diminishing and the General Motors stock, which would carry the company through later difficulties, was still suffering from the battering of the past year. Du Pont had already laid off two thirds of its work force in the immediate postwar retrenchment and further cutbacks would be difficult. The company had little to fear from the labor movement, which was virtually nonexistent in the industry, but reductions would damage production plans, particularly as they would have necessarily involved skilled workers and technicians. Attention therefore focused again on reorganization of Du Pont's marketing operation as the key to reviving profits.

Pickard and his allies resurrected the recommendations of the 1920 study group for reorganization of the company at the operating level along product lines in explosives, dyestuffs, pyralin products, paints and chemicals, and Fabrikoid, subject ony to standardized accounting and statistical reports under the Treasury Department. The Executive Committee would no longer have any direct responsibilities for line operation but would review the work of the line managers in terms of the earnings of their specific product lines, and would plan the future development of the company as a whole.

Although Irenee continued to express doubts about the degree of decentralization proposed, the crisis conditions of 1921 undermined a defense of the status quo. The fact that Pierre du Pont was implementing a similar program of decentralization at G.M. may also have influenced the situation, for Pierre remained the senior figure at Du Pont. At any rate, the company proceeded to adopt most of the recommendations for reorganization. After 1921 the general managers of the product line exercised considerable autonomy over purchasing, production, and sales, subject to intensive review of performance based on the return on investment standards developed earlier. The Executive Committee mapped the future direction of the company, allocated funds in keeping with these plans, and evaluated the performance of the divisions.[6]

The significance of this reorganization for Du Pont's recovery from the 1921 slump is not altogether clear. By 1925 company net earnings were back up to $24 million, with a rate of return on investment of 11.5 percent, the company's base target, and earnings soared further as the "dollar decade" peaked. In 1929 earnings were $78 million and return on investment was 15 percent. But American manufacturers as a whole performed exceptionally well in this period, so that the improved economic conditions account for part of this growth. Furthermore, Du Pont's earnings from its G.M. investment, a glamour stock of the 1920s, contributed over 50 percent of its net income in 1925–29, and subsidized the expansion of chemical lines such as dyestuffs and synthetic ammonia.[7] In the longer term, of course, the diversification policy paid off. Even in the bottom of the Great Depression in 1932, Du Pont's net earnings were $10 million and the chemical lines contributed over 50 percent of this total. The reorganization into multidivisional operating units on product lines provided Du Pont with continuing flexibility in moving into new fields, while the financial reviewing process ensured continuing coordination of investments and operations.

Other American chemical companies followed Du Pont along the diversification path in the 1920s, although none of them extended to such a wide variety of fields. Hercules and Atlas powder companies, the former adjuncts to Du Pont, diversified into nitrocellulose-related lines in this period. Hercules branched into lacquers, primarily for automobile coatings and resins derived from pine wood used in paints and varnishes, and initiated research in the 1920s on uses of turpentine that eventually led it into insecticides. Hercules also expanded its explosives capacity, absorbing the defunct Aetna Explosives in 1921 and a 50 percent share in the Mexican powder company in which Du Pont had invested earlier. Atlas diversified with less alacrity, concentrating more on research than mergers for developing new fields, but by the end of the 1920s, Atlas also had entered substantially into lacquers and industrial alcohols. Unlike Du Pont, both those companies focused on diversification in industrial supply fields rather than on products encompassing retail markets, so that the pressure for internal reorganization was not as great. Nevertheless, the two companies did move toward a multidivisional structure during this decade, reflecting the continuing Du Pont influence.[8]

Among the most active diversifiers in chemicals were American Cyanamid and Union Carbide. Both expanded largely through mergers and takeovers, following Du Pont's lead. American Cyanamid remained "intensely fertilizer conscious" up to the middle of the 1920s when its two most important figures, Frank Washburn and James B. Duke, died. Washburn's successor, William Bell, embarked on a major diversification drive, culminating in 1929 with the purchase of companies in plastics, dyestuffs and organics, and heavy chemicals. Its dyestuff acquisitions, Calco Chemical Co. and the formerly German-owned firm of Heller and Metz, made Cyanamid the third largest producer of organic chemicals in the United States in the interwar period. Union Carbide continued to expand both into chemicals and in other fields such as metal alloys, vanadium mining, and even hydroelectric power. The company moved carefully into chemicals, initiating research in acetylene in conjunction with the Mellon Institute in Pittsburgh during 1914–18. When the company was restructured along product lines in 1921, Union Carbide's chemical division focused on the development of aliphatic organic chemicals, a field separate from coal-tar based compounds which eventually led the company into chemicals derived from natural gas.[9] Although Carbide's chemical operations remained relatively small in terms of overall company sales in the 1920s, accounting for only 5 percent of net profits in 1929, the base was laid for a significant expansion in the following decade.

One exception to this pattern of diversification was Allied Chemical and Dye, which remained under the autocratic domination of Orlando Weber up to 1935. During the 1920s Weber concentrated on completing the vertical integration of the company, and on developing a giant synthetic ammonia plant in Hopewell, Virginia, to supplement the Syracuse plant. Weber eschewed moving out of the industrial supply fields, avoiding the organizational problems encountered by Du Pont and retiring Allied's large preferred stock issue and outstanding loans while financing plant expansion from earnings. Weber's conservative approach worked well in the 1920s. Allied stock sold high, buttressed by substantial dividends and rising sales volume in dyes and ammonia. The company's failure to diversify and develop a strong research base, however, ultimately would undermine Allied's position in the depression era when the bottom dropped out of the fertilizer market and other industrial customers, particularly in textiles, were hard hit.[10]

Companies outside the chemical industry also adopted Du Pont's multidivisional structure. General Motors developed its own variant at the same time as Du Pont, reflecting to some extent the cross-fertilization of ideas through Du Pont directors of G.M. Although the problem there was not too much centralization but too little, a balanced system of financial coordination and operational autonomy evolved in both companies. Over the next three decades a variety of other firms followed Du Pont's lead. Some, like U.S. Steel and U.S. Rubber, began moving toward a multidivisional coordinated structure in the late 1920s when Du Pont influence was directly involved, but the form was also embraced by companies unrelated to Du Pont, reflecting the impact of Du Pont's highly visible image in the interwar period as one of the most successful companies in the country.[11]

Expansion in the 1920s

The problems of reorganization in the 1921 depression temporarily halted Du Pont's growth and diversification, but within two years the Wilmington company had resumed its course of expansion by merger, supplemented by a practice particularly emphasized by Irenee du Pont, entering new markets by acquiring American patent rights to major products and processes. This was not a novel method for Du Pont. As early as 1890, it will be recalled, Alfred du Pont had purchased American rights to smokeless powder from a Belgium firm. But Irenee had been involved in patent negotiations from 1910 when, as head of the Development Department, he had participated in the acquisition of British artificial leather processes; and he had been a strong supporter of the move into dyestuffs through the Levinstein contract during World War I. The focus on expansion by patent acquisition was to be a hallmark of Irenee's years as president. This strategy would lead Du Pont into increasingly closer ties with foreign chemical firms in the 1920s.

In 1919 Irenee dispatched a mission to Europe to investigate the prospects of acquiring rights to viscose fiber from a French company, Comptoir des Textiles Artificiels, reflecting the new Du Pont head's continuing interest in that field which the Development Department had considered entering in 1912. While in France, the Americans discovered a new product, a transparent packing material made by La Cellophane, a subsidiary of Textiles Artificiels. Although further development was interrupted by Du Pont's internal problems, in 1923 it initiated joint ventures with the two French firms in artificial silk and cellophane, establishing adjacent plants at Buffalo, New York, the following year.[12]

Du Pont also extended its research efforts in paints and plastics. One of the major results of this work was the development of "Duco" lacquers for automobile finishes and "Dulux" enamels for house paints. One of the incentives behind Du Pont's acquisition of the Arlington Co. in 1915 had been that Arlington produced a nitrocellulose-based enamel that dried faster than the resins and varnishes commonly in use and could potentially be marketed to fit the streamlined mass-production schedules of the automobile industry. The major problem was that the nitrocellulose enamel was not tough enough to withstand hard outdoor use, limiting its value as an automobile finish. After several years of experimenting, E. M. Flaherty of Du Pont's Parlin, New Jersey, laboratory developed a suitably hard enamel that was gradually introduced on the market in small consumer items such as toys and brushes. In 1922, Du Pont induced Charles F. Kettering, General Motors' vice-president in charge of research, to try out the new lacquer. The energetic Kettering pushed the adoption of the Du Pont product, and a year later G.M. had begun using "Duco" finishes on its Oakland model. By the end of the decade almost every American auto manufacturer had followed suit. The major holdout, not surprisingly, was Henry Ford, who is supposed to have said of his Model T, "Any customer can have a car painted any color he wants so long as it is black." As the American auto companies expanded their manufacturing operations overseas, Du Pont followed, licensing its "Duco" process to European companies with the understanding that they would use the trade name.[13]

At the time when Du Pont was pushing G.M. to adapt its "Duco" finish, Kettering and his chief researcher Thomas Midgley were engagaged in research on chemical additives that would reduce "engine knock" in automobile fuels, a project they had begun even before Kettering's research firm, Delco, had merged with the auto company. By the end of 1921, Midgley had developed a compound, tetraethyl lead, that diminished the knock problem. Du Pont had also been carrying on research on antiknock compounds based on bromine derived from sea water and had spent about $50,000 on the undertaking when Kettering informed Irenee du Pont in 1922 of Midgley's accomplishments. Irenee and his chief chemical researchers, Dr. Charles Reese and Dr. Charles Stine, persuaded Kettering and Sloan to allow Du Pont to take over development of tetraethyl lead for both companies.

At this point, however, matters became more complicated. Standard Oil of New Jersey had been developing its own version of an antiknock compound based on chlorine rather than bromine, which was cheaper but produced a more unstable variant of tetraethyl lead. G.M. had drafted an agreement with Du Pont but delayed action while investigating the Standard Oil product. Du Pont had by this time established a small tetraethyl lead plant at Deepwater, New Jersey, and was alarmed at the emergence of so formidable a rival in the field. Throughout 1923–24 the three large firms bickered over the issue. Despite its close ties with Du Pont, G.M. did not want to antagonize Standard Oil, and Sloan proposed a joint venture with Standard to provide "a check on prices charged by du Pont" for tetraethyl lead. In August, 1924, a mutually satisfactory arrangement was concluded in which G.M. and Standard Oil established Ethyl Gasoline Co. that would market the knock-free fuel. Du Pont would provide most of the tetraethyl lead from its Deepwater plant, which was expanded to produce 700 gallons per day, while Standard would supply a smaller amount from its plant in Bayway, New Jersey. Du Pont had thus acquired the lion's share of the tetraethyl lead market as the only company licensed to produce it for Ethyl Gasoline Co., despite the fact that both G.M. and Standard Oil had pioneered in the field and developed the most efficient process, a tribute to Irenee's negotiating talents and Du Pont's influence over G.M.[14]

Du Pont's problems with tetraethyl lead did not end, however, with this corporate arrangement. In 1923 a Du Pont employee at Deepwater died, apparently from contact with the compound, and there were rumors that other workers had been hospitalized in Du Pont–owned clinics. In October, 1924, shortly after the conclusion of the Ethyl Gasoline arrangement, five workers at Standard's Bayway plant died and in 1925 the *Nation* charged that over 300 of the Deepwater employees had been poisoned and that Du Pont had suppressed the information for almost a year.

Du Pont had not been unaware of the potential health hazards involving tetraethyl lead. In 1922 Senator Coleman du Pont, mindful of the publicity problems that might ensue, recommended that the company investigate the risk of "lead poisoning in thickly populated centres from the exhaust of autos using gasoline doped with tetraethyl lead," and in 1923, after the first industrial illnesses surfaced, Du Pont cooperated with the U.S. Bureau of Mines in a study of the health risk of the compound. This investigation concluded that the risk was tolerable to permit continued production, but after the Bayway deaths, chemists from Harvard University criticized the report, fueling demands for further study of the environmental dangers, and local boards of health in New York, New Jersey, and Pennsylvania banned the use of ethyl gasoline. The Bayway and Deepwater plants were shut down in 1925 at the order of the New Jersey Labor Commissioner pending an investigation by the U.S. Surgeon General.

The Surgeon General's report in January, 1926, concluded that while tetraethyl lead did pose safety problems in production, the levels of toxic lead released by automobile exhaust would be tolerable provided the ethyl content of gasoline was limited. The Deepwater plant was reopened later that year after safety measures and improved ventilation were installed. Critics at the time argued that the report had been watered down through the political influence of Du Pont, and later studies contested the sanguine conclusions of the report on the effects of auto pollution, but for the time being the issue was buried, and Du Pont resumed its role as major supplier of tetraethyl lead to the American market. Du Pont's only serious competition in the field up to 1940 was Sun Oil, which developed an alternative antiknock compound marketed as "Blue Sunoco." Between 1924 and 1947, when the various patents to the processes expired, Du Pont is estimated to have had net earnings of $86 million in this field, an average of $3.7 million per year.[15]

Other technical developments exploited by Du Pont in the 1920s were less controversial but no less lucrative, particularly in plastics. Du Pont's foothold in this field through the Arlington Company had proved useful in the development of related fields such as cellophane and rayon, and in 1925 the Wilmington firm absorbed Viscoloid Company, manufacturers of small plastic products such as combs and toys, complementing Arlington's emphasis on industrial plastics and lacquer. The two companies were combined into Du Pont's Viscoloid division. Meanwhile, Du Pont had been investigating new markets for its pyroxylin plastic. In 1924, Du Pont invaded the photographic film field that had been dominated by Eastman Kodak, through a joint venture with the French film company, Pathe Exchange, Inc. Kodak retaliated with a foray into the cellophane market with its own variant of transparent wrapping material. Despite this counterthrust, Du Pont's rapidly growing cellophane sales were unaffected, and it continued to move into the cinematic film supply field, holding an estimated 40 percent of the market for negative film in 1924, and a smaller 20 percent share of the positive film market where the American producers all faced vigorous competition from the German Agfa-Ansco company.

Du Pont also moved, with more circumspection, into the glass industry, forming a joint venture with Pittsburgh Plate and Glass to manufacture shatterproof windows by combining sheets of pyroxylin with regular glass. The major competition in this field was the Celluloid Company which had been established in the 1880s as one of the earliest manufacturers of cellulose plastics, but had now fallen on hard times after failing to maintain its research and development efforts. In 1927, however, the Celluloid Co. was taken over by a more vigorous operator, Celanese Corporation, controlled by the Swiss brothers, Camille and Henri Dreyfus. The Dreyfus brothers had pioneered in the development of cellulose acetate film, which was less flammable than nitrated cellulose film, in the years before World War I. During the war they had built plants in England and the United States to supply cellulose acetate for airplane production. When the war ended and the Allied governments canceled their contracts, the Dreyfus brothers converted their plants to the manufacture of a synthetic fiber based on cellulose acetate, called "celanese," which they began to market in 1924–25. The Celanese Corporation not only produced yarn but also integrated into dyeing and weaving when American textile manufacturers balked at the price of cellulose acetate yarn, which was initially $1 per pound higher than viscose rayon. Imaginative marketing of Celanese's fine and richly dyed cloth and technical improvements that rapidly reduced the cost of production assured the Dreyfus brothers a firm foothold in the synthetic fabric field, and stimulated Du Pont and American Viscose to develop cellulose acetate yarns by the end of the 1920s. Meanwhile, Celanese Corp. branched not only into the safety-glass field but also into small plastic products through its acquisition of the Celluloid Co. In 1931 Du Pont sold its interest in the shatter-proof glass venture to P.P.G., although it continued to supply pyroxylin sheets, more or less conceding the market to Celanese while confining its plastic operations except for film primarily to industrial supplies.[16]

In all of these ventures Du Pont had developed fairly regular procedures for entering, exploiting, and then protecting markets based on methods evolved by the Development Department in the war period. When entering a new field Du Pont would identify the leading technical innovator, then approach that enterprise not only to acquire patent rights but also to procure assistance in every phase of production, usually through a joint venture arrangement in which Du Pont managers were quickly integrated with the partner firm. If the venture paid off, Du Pont would buy out its partner or at least extend its position to ensure a majority share of the business. Potential rivals with alternative technologies would also be enticed with proposals for joint ventures or license-exchange arrangements, and threatened with costly patent litigation if they rebuffed these blandishments, a practice that did not endear Du Pont to other businessmen and contributed to later antitrust difficulties with the government. As in the explosives field in the nineteenth century, however, Du Pont preferred market stability once its own position was firmly established.

Du Pont selected new production sites carefully, taking advantage of all the cost benefits provided by technical improvements in such fields as transportation and power generation. The company broke the trail for a number of industries centered in the Northeast, establishing three rayon mills and four cellophane plants as well as its large synthetic ammonia plant in the Southern states in the 1920s. In addition to the lower wage rates the South provided cheap hydroelectric power and low-priced real estate, and new factories were designed from the ground up to make full use of factory organization based on electricity.

In the distribution and sales of new chemicals, Du Pont adopted the German practice of hiring university graduates with some experience in the natural sciences, and established training programs to ensure that their sales and service people were familiar with the myriad technical details involving new products. Advertising became increasingly sophisticated: the company expanded and centralized its public relations and advertising operations, and drew on Batten, Barton, Durstin and Osborne of New York for additional advice, as in the adoption of the slogan "better things for better living through chemistry," which became Du Pont's slogan for more than three decades.[17]

Du Pont was not unique in these activities, but it was in the forefront of the American chemical industry in production and marketing innovation. To a large extent Du Pont's dominant position in U.S. chemicals at the end of the 1920s was the result of its massive financial resources accumulated from war production and G.M. earnings, and its deliberate strategy of market control where possible, rather than from its technical leadership. Nevertheless, Du Pont's achievements in organization and planning the exploitation of technologies acquired from others must be accorded recognition for contributing to the company's tremendous rate of growth in this era, averaging almost 10 percent a year over the decade, and 14 percent a year in 1925–29.[18]

Du Pont also continued its traditional practice of expansion by merger in the 1920s; toward the end of the decade the company seems to have been caught up in the bull market spirit and acquired large blocks of stock of other industrial giants, attracting the notice of the normally somnolent antitrust authorities. The buying spree tapered off in 1929, but even during the depression Du Pont picked up the occasional small firm while consolidating its position in fields such as ammonia, rayon, and cellophane by buying out partners.

One of the largest and most important of these acquisitions was the Grasselli Chemical Co. in 1928. Grasselli was one of the oldest diversified chemical companies in the country with assets of $56 million and common stock valued at between $117 and $145 per share in 1926. In addition to its sixteen operating plants producing various heavy chemicals, acids, and fertilizers, Grasselli had a subsidiary in explosives and another larger subsidiary in dyestuffs that it had acquired in World War I when it bought Bayer patents that had been seized by the U.S. government. In 1925 Grasselli had turned over sales rights to these dyes to General Dyestuffs Co., the new American subsidiary of the German firm, I. G. Farben, and subsequently the Germans took over all of Grasselli's dye operations: but even minus its dye facilities, Grasselli was a major factor in the American chemical industry and established in fields such as acids that Du Pont had never entered.[19]

In 1928 Walter Carpenter, Jr., of Du Pont proposed a merger with Grasselli involving a direct exchange of shares on the basis of one share of Du Pont common for five shares of Grasselli and an equal exchange of preferred stock. The Grasselli name was retained even after the heavy chemical operation was formally integrated into Du Pont's divisional organization. Du Pont made no effort to buy the Grasselli dye works, on the grounds that this acquisition might raise legal problems under the Clayton Act even though Grasselli was recognized as "one of our strongest competitors" in dyestuffs. While Du Pont may have anticipated antitrust difficulties, the decision not to pursue Grasselli Dyestuffs was probably influenced more by Du Pont's intricate and evolving relationship with I. G. Farben. At any rate, the cost to Du Pont of the Grasselli transaction was $64.5 million, for which it received in return a significant position in the heavy chemicals industry, complementing its growing role in the organic chemicals field.[20]

At the same time that Du Pont was contemplating the Grasselli acquisition, more grandiose ventures were afoot. In February, 1927, Irenee du Pont asked Carpenter for a projection of stock prices of U.S. Steel. Carpenter noted that the steel company's stock appeared undervalued and was likely to rise. Three months later Irenee authorized purchase of 114,000 shares of U.S. Steel common at $122 per share, and a month later Du Pont joined Koppers Co. of Pittsburgh in further purchases.

The acquisition of a $14 million block of U.S. Steel in such a short time aroused the attention of both the New York financial community and antitrust lawyers in Washington. In response to a sharp inquiry from Assistant Attorney General William Donovan, Carpenter maintained that the block of shares involved constituted less than 2 percent of U.S. Steel's voting stock and Du Pont's sales to U.S. Steel were less than 1 percent of that company's total purchases so that no "community of interest" could be discerned. Despite this disclaimer, the government kept up the pressure. The Federal Trade Commission initiated an investigation over the protests of its own chairman, William Humphrey, who characterized the affair as an "illustration of bureaucracy gone insane." In the early months of 1928, Du Pont sold its U.S. Steel stock, netting $2.3 million. The company retained $660,000, providing a 6 percent return on its investment.

Du Pont's intentions in this episode are not altogether clear. Irenee argued that it was simply looking for an outlet for investing surplus earnings, a view shared by F.T.C. chairman Humphrey. Others discussed more devious aims. The F.T.C. majority, like Donovan at Justice, noted the Du Pont–G.M. connection and considered the prospects of an alliance between three firms so closely interrelated in production, making clear its suspicion of the G.M. relationship. There was also speculation that Du Pont might be contemplating a candidate of its own choice at the head of the steel firm: the stock purchases occurred fortuitously close to the death of U.S. Steel's chairman Elbert Gary, and there were rumors of rivalry between Du Pont and J. P. Morgan over the succession, although this argument required the improbable assumption that the New York bank and Wilmington chemical firm were on bad terms, which does not seem to have been the case.[21] Du Pont's own position on this issue was that it anticipated a more "aggressive" leadership after Gary and expected to make a quick profit, as indeed it did. It is interesting to note that Myron C. Taylor, who became U.S. Steel's chief operating executive in 1927, introduced a variety of organizational changes patterned after Du Pont and G.M.

Du Pont's penetration of U.S. Rubber was more gradual, apparently more deliberately conceived, and ultimately more enduring than the U.S. Steel foray. The rubber giant, formed in 1892, had fallen into a sad state by the mid 1920s. Price-cutting competitors like Firestone had eroded its market share, its factories were outdated, and it was carrying a debt load requiring more than $6 million a year in interest charges alone. The industry as a whole, however, was booming as a British-sponsored rubber cartel effectively restricted supplies, driving natural rubber imports to over $1.25 per pound in the U.S. market, aggravated by the rising demand for rubber by the auto industry. To overcome the shortfall, American companies such as Firestone had established new rubber plantations in Africa, and Ford had integrated into rubber production in 1928. Meanwhile, chemical firms including I. G. Farben in Germany and Du Pont and Standard Oil (New Jersey) in the United States had begun research on synthetic substitutes.[22]

In 1927–28 crude rubber prices began to fall as the British succumbed to foreign pressures and the growth of new competition, and announced the termination of the cartel in 1929. Weak overcapitalized companies such as U.S. Rubber and its traditional rival, Goodyear, faced even more severe difficulties, but some observers, including Irenee du Pont, foresaw eventual market recovery in which "efficient and well managed companies" could anticipate "a reasonable profit." In the summer of 1927 Irenee and a private syndicate that included other Du Pont executives began buying U.S. Rubber shares, joined by Pierre du Pont in 1928, by which time the group had accumulated over 200,000 shares, controlling about 18 percent of the rubber company's common stock and 11 percent of its preferred stock. According to Irenee du Pont this undertaking had begun as simply, if improbably, a "speculative investment," but subsequently he was approached by Edgar Davis, who held large blocks of stock in both U.S. Rubber and Goodyear, and the New York investment banker Clarence Dillon, with a proposal to merge the two companies, which together held a 50 percent share of the American rubber market and to impose "stronger and more capable management at the top." Sloan of G.M. was also brought into the discussions but declined to enter an arrangement in which his company would not "own the facilities outright."

Undeterred, Irenee took the proposal to his own company. Walter Carpenter endorsed the idea of a merger but advised against a direct Du Pont involvement as at best "we could have but a minority participation" in the consolidated rubber company. The merger in fact did not occur, possibly because of Du Pont's caution. Nevertheless, the Du Pont corporation supported the investment by Irenee's syndicate in U.S. Rubber, not only because of that company's role as a supplier to the auto industry but also because U.S. Rubber had begun to diversify into chemicals in the 1920s. With their de facto 30 percent control of U.S. Rubber, Irenee and his partners arranged for the reorganization of the company at the end of 1928, installing a Du Pont company executive, Francis B. Davis, as president of U.S. Rubber and introducing a multidivisional structure along Du Pont lines. Du Pont and U.S. Rubber later cooperated in developing the synthetic rubber, neoprene, and in 1931 G.M. entered a contract with U.S. Rubber for the supply of 50 percent of its tires, an arrangement that continued for thirty years.[23]

The forays into the steel and rubber industries marked the high point of Du Pont's expansion in the 1920s. The company continued to pick up minor acquisitions, even after the depression began, most notably the Roessler-Hasslacher Chemical Company, that specialized in insecticides, in 1930, and the financially troubled Remington Arms Company, in which Du Pont bought majority control in 1933. By this time, however, Du Pont had begun to shift from a strategy of expansion by merger and purchase of foreign technology to an approach that emphasized development from its own formidable research operations. Du Pont's growth in the booming twenties was impressive: assets, not including the lucrative G.M. investment, more than doubled between 1919 and 1929. Net earnings had quadrupled in the period, a remarkable performance in view of the fact that in 1919 Du Pont was still profiting from wartime munitions sales. The surplus account had doubled as well, despite the large outlays for acquisitions. Although the depression of the 1930s would erode Du Pont's earnings, the company's scale of operations, command of technology, and the organizational innovations adopted in 1921 would carry it through the worst period in America's economic history in the twentieth century. At the end of the depression Du Pont would remain the supreme chemical manufacturer in the United States and one of the most technically innovative companies in the country.

CHAPTER SEVEN

The Consolidation of the International Chemical Industry

The decade of the 1920s marks a watershed in the development of the international industry. The momentum of technological innovation and commercial development initiated in the war period continued as new uses of chemical products were discovered and new markets exploited. The 1920s also witnessed the emergence of consolidated, vertically integrated corporations in the field, again continuing a trend the war had set in motion. This process of concentration was most marked in continental Europe, particularly in Germany, where the wartime alliance of eight of the largest chemical firms culminated in the formation of the I.G. Farben Industries A. G. in 1926 after more than eight years of negotiating among its members. The leading Swiss chemical and dye firms also amalgamated along the same lines as the Germans after the war, although the degree of integration was less formal. In Italy Montecatini expanded by merger with a range of diversified chemical companies and became the major firm in the field by the end of the decade, although its business was dwarfed by its counterparts in the North. In Britain, Nobel and Brunner-Mond merged with smaller firms in 1926–27, partly in response to the formation of I. G. Farben, achieving a scale and degree of diversification in British Empire markets roughly approximating that of Du Pont in the United States.[1]

At the same time the heavy capital costs and increased scale of operations of these giant enterprises created serious problems for them. L. F. Haber has observed that most of the European "chemical concerns were usually production—rather than market—oriented" and the stimulus for large-scale organization did not necessarily reflect growing demand for chemical products.[2] The initial moves toward cartelization that preceded more formal amalgamation of firms were occasioned in part by the problems of sluggish markets in the immediate aftermath of the war, and chemical manufacturers continued to face difficulties adjusting production to cyclical changes in demand. The steady deterioration of agricultural commodity prices in the late 1920s affected the demand for synthetic fertilizer on which both I. G. Farben and I.C.I. had pinned high hopes, and the industrial depression after 1929, accompanied by increasingly restrictive trade measures imposed by the industrial nations, presented equally severe challenges in the other chemical markets.

As in the past, chemical manufacturers turned to the familiar devices of price restriction and informal market allocation to meet the crisis. In the waning years of the "Prosperity Decade" negotiations among what Harry McGowan of I.C.I. called the "great powers of the chemical manufacture of the world" became increasingly complex and crucial to the survival and growth of these firms. Each of the major integrated chemical enterprises—I. G. Farben, I.C.I., Du Pont—developed a "foreign policy" that encompassed a range of mutual problems, including not only market restraints but also technological exchanges, joint ventures, intercompany investments, and related matters. Each of these companies also established administrative departments to monitor negotiations and implementation of agreements, the functional equivalents of foreign relations departments in national states.

As this description suggests, the elaborate policy and organizational innovations that evolved among the major chemical firms in the 1920s reflected more than a short-term response to unstable or declining markets. In the nineteenth century the predecessor companies of these integrated concerns had consistently sought to establish stability in their home markets through agreements with competitors not only to ensure a steady rate of return on investment but also to provide mutual access to new technological developments. In the decades preceding World War I these arrangements extended into international markets: Du Pont and Nobel in explosives and the Solvay group in alkalis maintained loosely built but enduring alliances across borders. In the organic chemicals field, the Germans and Swiss dominated international empires. The war of 1914–18 disrupted these systems and created a new situation of international competition, particularly in dyes and organic chemicals, but did not eradicate the preference of chemical manufacturers for stability and mutual accommodation. The British were the most outspoken advocates of this viewpoint. Harry McGowan of Nobel described the formation of the British chemical giant, Imperial Chemical Industries Ltd., as "the first step in a comprehensive scheme . . . to rationalize the chemical manufacture of the world" through a network of alliances among international producers.[3] The founders of I. G. Farben were more interested in reestablishing the prewar position of

German domination, but they too were prepared to negotiate international agreements as a means to this end. The other European chemical producers were, for the most part, too small to do more than accept junior roles in the arrangements made by the Germans and British.

Du Pont's position in this emerging postwar system of chemical alliances was ambiguous. As in the prewar period, the company was wary of any direct involvement with foreign firms that would conflict with American antitrust laws. Since 1907 it had deliberately avoided substantial direct investment abroad, except for the nitrate fields in Chile and the joint venture with Nobel in Canada, and kept out of export markets as well. A premature entry into the international dye market in the early 1920s confirmed the predisposition. The consistent aim remained protection of the American domestic market. At the same time, Du Pont wanted access to foreign technical innovations, if only to protect its domestic position, and some kind of understanding with the international companies was necessary to stem any foreign forays into the large and tempting American market. These considerations determined Du Pont's stance in the elaborate negotiations of the 1920s.

Du Pont's strategy for dealing with potential foreign invaders was twofold. On the one hand, the company turned to the traditional device of American industrialists, the protective tariff, to head off competition particularly in the new and precarious product lines such as dyestuffs. During 1919–22 Du Pont lobbyists, together with other representatives of the American chemical industry, established an "American sales price system" on imported chemicals that was to endure for the next forty years. At the same time Du Pont began talks with foreign manufacturers, especially the Germans, with the aim of promoting technical exchanges that would incidentally lead to a broader understanding about the informal division of world markets.

With its own market protected on one flank by tariffs, Du Pont sought by these agreements to avoid penetration by other methods, either through direct investment or unforeseen technological breakthroughs. Equally important, Du Pont wanted access to the technology of foreign producers to improve its own products.

This essentially defensive viewpoint was shared by the British, and the prewar Du Pont–Nobel system of technical agreements was rapidly resurrected and refined as Nobel was transformed by mergers into a diversified chemical firm parallel in many ways to Du Pont. The Anglo-American negotiations were not without friction on particular points, but a general accord prevailed. Relations with the Germans were never as cordial. After driving the Americans out of the international dye market and isolating the British by the mid-1920s, I. G. Farben posed a threat to Du Pont within its own domain through various American affiliates. After several years of acrimonious bargaining an uneasy detente prevailed as the depression endangered all of the chemical giants.

In certain respects the alignments of the "great chemical powers" in the period 1919–39 paralleled those of the international political system on the eve of World War II. Except for minor forays abroad in various product lines, the Americans were preoccupied with domestic matters, but were increasingly drawn into a defensive posture with the British culminating in 1928–29 with what would later be called the "Grand Alliance," which in effect was intended to counter the aggressive commercial policies of the Germans even though both Du Pont and I.C.I., along with numerous other smaller firms, maintained a broad range of agreements with I. G. Farben. Although the corporate diplomats of this era proved to be far more adept at reconciling differences among themselves than did their counterparts in governments, the similarities in circumstances are quite remarkable. This situation was not unique: the international steel cartel of the late 1930s also reflected this balance of power between the export-oriented Germans and the more conservative American and British steel makers.[4] In other industries, such as electrical equipment, automotives, and oil, a different alignment of interests existed. A common feature of this interwar period, however, was the presence of an elaborate system of commercial relationships in the international economy that transcended the political system of nation states and to a certain extent can be seen as the precursor of the system of multinational enterprises that emerged in the aftermath of World War II.

Consolidation in the European Chemical Industry

Although the final steps in the consolidation of the large integrated chemical enterprises in Europe did not occur until 1925–26, the process began during World War I and was in large measure the result of the demands imposed by belligerent governments on their chemical industries. In Germany the formation of I. G. Farben represented the culmination of trends toward amalgamation that began in the 1890s, but the war undoubtedly hastened developments by promoting a tremendous increase in the output of the industry and encouraging a greater cooperation among firms, a trend that was augmented further by the instability of postwar markets.

In the decade prior to the war, as recounted earlier, the major German chemical firms had formed two large cartels, the Hoechst-Cassella "community of interest" (I.G.) and the Bayer-B.A.S.F.-Agfa "Dreibund." The two groups entered market agreements, particularly for exports, but otherwise remained separate up to 1916. When war broke out and the export market abruptly collapsed, Carl Duisberg of Bayer resumed his long-standing effort to integrate the two groups. His pleas were now endorsed by the German government, which abandoned its previously suspicious attitude toward the monopolistic implications of centralization in the chemical industry. In 1915–16 the two groups negotiated the formation of a new organization, the Interessengemeinschaft der deutschen Teerfarbenfabriken, that also included a smaller dye works, Weiler-Ter Meer, and Griesheim Elektron, a company that had pioneered in the refining of light metals such as aluminum and magnesium through electrolytic processes.

The new I.G. was organized along lines similar to those of the two predecessor groupings. Profits were pooled and redistributed, with Bayer, B.A.S.F., and Hoechst taking the largest shares. The member companies agreed to concentrate production in the most efficient plants, supplying the others at cost plus 10 percent, although this objective proved hard to achieve in practice. Research costs and management salaries were centralized. But the member firms retained their independence in operations, despite vigorous demands by Duisberg for integration similar to the Americans'. The executives of each of the firms were reluctant to surrender "their own personal power" through "the transfer of policy making decisions to a central board of directors where their voice would be only one of many."[5] Furthermore, certain key technologies were not made available to the organization, including Griesheim Elektron's electrolytic processing of metals and B.A.S.F.'s nitrogen fixation processes, although by the end of 1916 B.A.S.F. relinquished its synthetic ammonia know-how to the I.G. in return for new capital to expand production.

The I.G. proceeded to absorb other small dye works in Germany and initiated research in areas that were to become extremely important in the postwar era: the production of a synthetic gasoline from coal through high-pressure synthesis similar to the Haber-Bosch process for synthetic ammonia; and the development of a synthetic rubber. The natural sources of both these materials were in short supply in blockaded Germany and B.A.S.F. was rapidly developing its engineering capabilities in the high-pressure synthesis area, but the war ended before either of these developments had reached the point of substantial industrial production.[6]

The postwar era posed serious problems for the I.G. The Germans had lost their monopoly of the world dyes although the I.G. had the technical capabilities and every intention of recovering the market. The Allies, particularly the French, seemed intent on acquiring access to the synthetic ammonia process, or, failing that, the destruction of B.A.S.F.'s Leuna and Oppau nitrogen plants. German chemical patents and plants in Britain, France, and the United

States had been confiscated and were already in the process of being redistributed to potential foreign competitors.

B.A.S.F.'s new chief executive, Carl Bosch, circumvented the French government's demands for the closure of the synthetic ammonia plants by arranging to share dyestuff technology with them, with I.G. retaining a 50 percent interest in its former dye manufacturing facilities in France. He was less successful in reacquiring other confiscated foreign patents and plants, however, and problems with the French resurfaced in the "Ruhr crisis" in 1923 when the French army seized the Ludwigshafen dye plant and the Oppau nitrogen plant after Germany refused to pay reparations imposed by the Versailles treaty. The plants were returned after the reparations issue was resolved (more or less) later that year, but in the meantime the French government had sold its interest in the French dye works to Kuhlmann, a large private concern that terminated the 1919 agreement with I.G. on the grounds that the Germans had failed to provide dyestuffs required under that contract during the period when its plants were closed down.[7]

Meanwhile, the German company labored to recover its prewar market. Ironically, the reparation provision of the Versailles treaty helped it achieve this aim. Under the treaty a large block of dyes, intermediates, and drugs were taken over by the Allies for resale to dye users at prices fixed by the Reparation Commission. But much of this stock of dyes turned out to be of inferior quality and the Germans were able to sell their high-quality lines on favorable terms. Furthermore, the sale of these confiscated dyes, while not constituting a particularly large block (about 20,000 tons, or 60 percent of the amount that the Allies initially announced), contributed to an existing market surplus in the sharp postwar recession, damaging the position of new dye manufacturers outside Germany, most of whom could only produce lower-grade products. As their sales fell off, these companies were unable to raise the new capital required to improve their technical capabilities, so that when the dye market began to pick up in 1923–24 the Germans were in the best position to take advantage of the situation.[8]

Nevertheless, the view from I.G. headquarters was anything but serene. In 1924 the Germans had recovered only about 50 percent of their prewar export markets, and the combination of new tariff laws in the United States and elsewhere and the shutdown of the largest plants in the 1923 Ruhr crisis had given foreign competitors a fresh start. Furthermore, a mysterious explosion in 1921 had devastated B.A.S.F.'s synthetic ammonia plant at Oppau, killing over 500 workers and injuring thousands more. The plant had been rapidly rebuilt, but the capital costs were substantial. The emerging technically oriented leaders of the I.G. such as Bosch and his proteges Heinrich Schmitz and Carl Krauch, who had supervised reconstruction at Oppau, contemplated even larger investments to develop synthetic fuels and reduce Germany's dependence on imported oil as well as diversifying I.G. out of the shrinking dyestuff field. The chemical companies had survived the inflationary chaos of 1923 intact, but the prospect of future financial instability remained.

All these considerations generated a new interest in corporate reorganization. As in the past the initiative was taken by Duisberg who proposed two alternatives to the existing structure of I.G.: a holding-company organization in which the member firms would retain formal independence while coordinating sales and production more closely than at present; or a "fusion" of the companies into a single unit. Somewhat surprisingly Duisberg, who had advocated a high degree of centralization in 1903, now endorsed the looser holding company form, reflecting perhaps the diminished position of Bayer in I.G. Carl Bosch of B.A.S.F. urged "fusion," influenced in part by his recent travels in the United States where he had observed the operations of integrated firms such as Du Pont and Standard Oil (New Jersey). At this time many German industrialists had become enamored of "rationalization" along what were conceived to be the lines of American corporations, offering the presumed advantages of increased efficiency and economies of scale in production, and also more immediate financial benefits from new issues of securities.[9]

During 1924–25 Duisberg and Bosch and their respective allies in I.G. debated the merits not only of centralization but also the degree to which the new organization should operate under established long-range planning as opposed to a more flexible and autonomous role for managers, similar to the controversies within Du Pont in 1920–21. At length the deadlock was broken through the efforts of Franz Oppenheim of Agfa who proposed a compromise in which product lines were organized into divisions with each product manufactured in at least two places to maintain comparative costs. Dye sales were concentrated in one office, which eventually took over responsibility for export sales of other products as the dye market declined. The members of I.G. all exchanged shares with the new corporation which was capitalized at $262 million, employed 114,000 people, and controlled about one third of the total output of chemicals and almost 50 percent of the total invested capital in the industry in Germany.

One final point remained to be resolved. Bosch wanted to change the name of the new organization to reflect its status as an integrated corporation. Duisberg, however, prevailed in this instance, pointing out that the I.G. name was well established in the export markets and it would be commercially unwise to jettison it. Thus the name I.G. Farbenindustries A.G. was retained even though the company operated for the next two decades as a single unit.[10]

Duisberg moved to the chairmanship of the board, leaving Bosch and his contemporaries in control of management.[11] These men were all exceptionally tough-minded executives in an industry whose leaders were not noted for sentimentality. Paternalistic toward their workers, many of the I. G. Farben chiefs, including Krauch and Schmitz, would later be convicted at Nuremberg of war crimes. These included maltreatment of slave laborers imported by the Nazi regime to their factories, notably the notorious synthetic rubber plant at Auschwitz, and plundering the properties of former competitors in countries occupied by the Third Reich during World War II.

The I. G. Farben men also earned a reputation among their business competitors as sharp negotiators. Confident of their technical superiority, they could be abrasive in personal encounters, yielding little while demanding much in return. Yet the Germans, no less than the British and American chemical manufacturers, ultimately desired to establish conditions of market stability, particularly when the depression diminished sales of dyes and synthetic fertilizers, and threatened their ambitious expansion plans in synthetic fuels and related products. Later in the 1930s as I. G. Farben became tied to the German military for subsidies and contracts, its "foreign policy" emphasized access to foreign technology. This dependence served the company well in the prewar period but also contributed to deteriorating relations with foreign firms, particularly Du Pont, which had always been suspicious of the Germans, and laid the groundwork for an unprecedented effort to dismantle the German chemical colossus in the aftermath of World War II.[12]

The consolidation of I. G. Farben in 1925–26 helped speed up the move toward amalgamation of British chemical manufacturers that culminated in the establishment of Imperial Chemical Industries in December, 1926, although this development was also the result of efforts that had begun during World War I. I.C.I. was formed through the merger of Nobel Industries, which had begun diversifying out of explosives along lines similar to, although not duplicating that, of Du Pont in the same period; Brunner-Mond, the large chemical manufacturer and its decrepit competitor, United Alkali Co.; and British Dyestuffs Corporation, the government-supported child of the war that had suffered constant setbacks since its inception in 1919 as the I.G. steadily penetrated its potential market.

The initiative for consolidation came from Nobel Industries, and more specifically from its chairman and chief managing director, Sir Harry McGowan, a man of florid countenance, opulent tastes, dictatorial tendencies, and far-reaching ambitions. During the war McGowan, sharing the Du Pont's doubts about the long-term prospects of the explosives industry, determined to apply the earnings of Nobel Explosives to diversification. The first step in this strategy was the merging of the various British explosives manufacturers within a holding company, Explosives Trades Ltd., established in 1918 with McGowan as its chairman, which incidentally enabled him to outmaneuver a potential rival at Nobel, Harold J. Mitchell, who remained a threat to McGowan's supremacy, however, and briefly superseded him as head of I.C.I. in the 1930s.[13]

Meanwhile McGowan laid the groundwork for future growth by investing in Levinstein Ltd., the dyestuff manufacturer which had taken over the Bayer factories in England, which gave Explosives Trades Ltd. a foothold in British Dyestuffs Corporation. In 1920 Explosives Trades was renamed Nobel Industries Ltd. with a capital investment of $58 million in explosives and dyes and an additional $14 million available for "other industrial investments." Under McGowan, Nobel emphasized "the management of money rather than . . . technological development" in any one direction. Over the next five years McGowan shifted over 60 percent of Nobel's capital out of explosives into new areas, including not only dyes, but also artificial silk (through a minor investment in British Celanese Ltd.), and especially the automotive industry, through purchase of shares in the Dunlop Rubber Co. and in General Motors (discussed earlier).[14]

Nobel's links with Du Pont extended well beyond the initial investment in G.M. McGowan was impressed by Du Pont's organization as well as its wartime diversification strategy, which he had followed. During 1920–21, Nobel Industries was reorganized along Du Pont lines, reducing the autonomy of independent boards that had functioned under Explosives Trades Ltd., centralizing purchases and sales, establishing a Development Department, all under the general coordination of Management and Finance committees, which, however, in practice were kept unde McGowan's control as the chairman of both groups. Meanwhile, McGowan undertook to rebuild the shattered prewar alliance with Du Pont. In 1919 he and Mitchell journeyed to Wilmington, ostensibly to observe Du Pont's technical advances, but also to discuss a new general arrangement. McGowan seems to have hoped to negotiate a mutual exchange of shares between Nobel and Du Pont, but the cautious Du Pont heads, mindful of their recent hard-won control of their firm, forestalled this proposal, as they were later to rebuff the Germans embarked on a similar quest. In 1920, Du Pont and Nobel entered a new "patent and process agreement" which arranged for exchanges of information in lines where the two companies had converging interests and provided for future discussions to bring new products under the agreement. The Americans were, as in prewar negotiations, reluctant to put in writing any statement that could be interpreted as restricting markets, and the 1920 agreement specifically provided that "where no patented or secret process is involved . . . the whole of the World is open to either party. . . ." But Pierre du Pont assured the British that Du Pont "would not seek to take from their competitors trade which had in the past been held by the latter, realizing that this would inevitably lead to retaliation."[15]

Nobel supplemented this arrangement with other foreign agreements, including a complicated deal with its former associate in the German explosives, that also involved Du Pont and is discussed further below. Its foreign flanks secured, at least in some fields, Nobel began developing new product lines in automobile metals, carburetors, artificial leather, and paints, all on a relatively small scale. Between 1917 and 1926 Nobel could boast net profits of $59.6 million, with explosives sales, particularly in 1917–19, accounting for 75 percent of the earnings.[16]

All was not well, however, in the Nobel empire. In 1925 net earnings from the new product lines was only about $1.1 million and some fields, like cellulose finishes, had yet to show profits. More troubling were the fortunes of Nobel's investment in British Dyestuffs. In December, 1919, shortly after the formation of the company the government's proclamation of an embargo on imported dyes was nullified by a court order, and a new law that restricted imports through a licensing system did not come into effect until January, 1921. In the interim German dyes (including reparation deliveries) poured into the British markets, principally in the aniline colors where British Dyestuffs' technical capabilities were concentrated. Although in 1921 laws protected the British market, British Dyestuffs' hope to penetrate the large foreign markets grew fainter. The 1916 arrangement between Du Pont and Levinstein, now a member of British Dyestuffs, had dissolved amid litigation. In a 1920 meeting between representatives of the British firm and Du Pont men, the lawsuit was settled and arrangements made to establish a new American outlet that would bypass the pending protective tariff legislation; but a final agreement on this latter point was delayed, since the British dyemakers hoped to negotiate a deal with the Germans and did not want to jeopardize this prospect.

Meanwhile, British Dyestuffs' sales plummeted steadily, stabilizing in 1923 at half than 1920 level. Even in its protected home market, the British company faced competition from the large blocks of imported dyes stockpiles in 1920–21, and it lacked the capital resources to maintain and expand dye technology. Furthermore, there was continuous bickering among the former heads of the constituent firms until they were finally removed following a report to the board of directors that criticized both for their failure to cooperate and develop more cost-efficient organization. Even McGowan seems to have given up hope for the venture, resigning his seat on the board in 1921, although Nobel retained its investment in the firm.

The reorganized board under Sir William Alexander, a veteran of the Ministry of Munitions in the war, sought to hold the line by negotiating with the Germans. The I.G. however, took full advantage of the well publicized weaknesses of the British company, demanding that British Dyestuffs restrict its sales to the Empire markets, and share 50 percent of net profits in return for providing technical and manufacturing information. British Dyestuffs users criticized the proposed agreement, fearing that they would be stuck with higher priced products inferior to those of direct customers of I.G. The British government refused to endorse the agreement and it lapsed at the end of 1924. Alexander was ousted and the British government shortly thereafter pulled out of the enterprise altogether. In its first seven years British Dyestuffs, following losses of more than $3 million, earned a total of $1.3 million, an average return on investment of 3.2 percent.[17]

Alexander's successor at British Dyestuffs, Reginald McKenna, approached McGowan in January, 1926, with a proposal that Nobel absorb the troubled firm. McGowan was reluctant to take on this sickly adolescent, but his interest was piqued by the prospect of creating a British counterweight to the newly consolidated I. G. Farben, an integrated firm that could hold British Empire markets against the foreigners. To accomplish this aim, however, Nobel needed to merge with a heavy chemical firm. More important, in terms of Nobel's own product lines, in explosives and nitrocellulose-based chemicals, access to a secure nitrogen supply was crucial. Patriotism combined with self-interest in directing McGowan's attention to Brunner-Mond.

That company had continued to dominate the alkali field in Britain as its chief prewar competitor, United Alkali, steadily gave ground. In 1916 Brunner-Mond invested in Castner-Kellner, thus encroaching on United Alkali's only major remaining source of revenues, the chlorine field, as well as giving it access to the electrolytic processes in alkali production. After the war, Brunner-Mond rebounded rapidly from the slump, thanks to plant expansion that had been initiated before 1914. The most important of Brunner-Mond's activities in the 1920s, however, focused on the development of synthetic ammonia.

In 1918 the British Ministry of Munitions bought a site at Billingham on the Tees River in northeast England, adjacent to coal, steel, and surplus electric power supplies, to build a nitrogen fixation plant. When the war ended, the government cast about for private investors interested in participating in the project. Brunner-Mond was approached, and sent technical observers with the British mission to examine B.A.S.F.'s Oppau plant in 1919. The most important of these observers was Lieutenant Colonel George P. Pollitt, who also became a director of the alkali firm in 1919 and used this position to lobby for an investment in the ammonia process venture. In 1920 Brunner-Mond agreed to establish a subsidiary, Synthetic Ammonia and Nitrates Ltd., with an initial capital of $6.8 million and access, through the Ministry of Munitions, to German patents.

Brunner-Mond was too small to finance such an undertaking on its own, and approached Nobel. McGowan was initially interested, but then faced a cash problem as his commitments to Du Pont for G.M. stock purchases fell due, and Nobel pulled out at the last moment. But Brunner-Mond, possibly entranced by the same visions of fertilizer markets that influenced B.A.S.F.'s Bosch, decided to push on alone. Between 1920 and 1925 the company invested almost $15 million in the enterprise, financed partly through the sale of its subsidiaries in the soap business, Crosfields and Gossages, to Lever Brothers. In 1922, the Billingham venture received a further shot in the arm when Ludwig Mond's son, Alfred returned to active direction of the company after his political career foundered. As expansive of vision as McGowan, Mond was impressed at the scale of the project and threw his weight behind it. His influence was to grow in the company as the existing managers proved unable to cope with Brunner-Mond's increasing difficulties.

Some of these problems related to Brunner-Mond's complicated dealings with Lever, who in 1924 brought a suit against the alkali company for violating a 1919 contract. The case ended badly for Brunner-Mond, which had to pay $5 million in damages and led to the resignation of many of the directors, and the suicide of the former chairman. Mond took over the job in early 1926.

There were other, equally complex issues looming. Relations with the aggressive Orlando Weber of Allied Chemical in the United States had deteriorated. Allied had agreed in 1921 to observe the prewar division of markets in alkalis, but Weber pressed for concessions on exports and the Solvay Process Co., an Allied subsidiary, joined the American export association, Alkasso, that was probing foreign markets.[18] Allied had also established a foothold in the synthetic ammonia field and could be a potential competitor to Brunner-Mond, although Weber seemed more intent on exploiting the American market. Du Pont was moving into the nitrogen field but had not yet developed a satisfactory process.

While McGowan's natural habitat was the British Empire, Alfred Mond took a more cosmopolitan view of the future. Having developed its synthetic ammonia capacity at great cost, Brunner-Mond was interested in the synthetic fuel development in Germany, and for this purpose an alliance with I.G. Farben looked more promising than a merger with Nobel. Discussions between the two British firms in the early summer of 1926 were inconclusive, and shortly thereafter Mond met with Bosch and the Solvays in Brussels. All parties agreed, inaccurately, that synthetic nitrogen fertilizers had a great future and proceeded to draw up plans for a "British I.G." to be dominated by Brunner-Mond that would also move into the hydrogenation field and would incidentally absorb British Dyestuffs. Since Allied Chemical was established in both dyestuffs and synthetic ammonia, and was linked to Brunner-Mond through Solvay Process Corp., it too should be brought into this general arrangement.

Unfortunately, when Mond went to the United States to negotiate with Allied, he "found Weber at his most truculent," not only rejecting the proposed role in the British I.G. but also proposing to terminate the 1921 agreement on alkali exports. Meanwhile McGowan had thoughtfully arranged to journey to America at the same time and book passage on the *Aquitania* for the return trip, the same ship the Mond party was using. Alfred Mond, dejected over the failure of his mission, was now more than willing to listen to McGowan's alternative plans, and a deal was struck en route that produced, before the end of the year, the Imperial Chemical Industries. The junior partners, British Dyestuffs and United Alkali, were rapidly bullied into the merger, and the new company was unveiled amid patriotic rhetoric about developing the chemical industry through the Empire.

The company was capitalized, somewhat excessively, at $275 million, roughly equal to I. G. Farben, although more than 30 percent of the purported value was in the form of "good will" rather than tangible assets. In certain other respects, I.C.I. resembled I.G. Farben, as both were involved in synthetic ammonia, dyes and heavy chemicals, but I.C.I.'s dyestuff output and market share was small compared to the I.G. and its technical position in the new fields far less advanced, a circumstance that would affect the subsequent negotiations between the European giants. I.C.I. was also involved in certain product lines such as automotive parts and materials that I. G. Farben had ignored, and which encouraged a greater interest, on I.C.I.'s part, on the activities of Du Pont.[19]

In the years to come, I.G. Farben would continue to set the pace in the development of new chemical products such as plastics and synthetic rubber and push aggressively into new markets. I.C.I. followed a more cautious path, at least prior to World War II, focusing on established fields, acquiring technology through cartel agreements, and seeking market stability that would protect its British Empire markets.

CHAPTER EIGHT

Du Pont's Foreign Relations: The German Stalemate, 1919–26

During the 1920s, Du Pont's foreign contacts became increasingly complex and wide-ranging, including a brief entry into the export market in Asia, joint ventures with French firms in the United States, parallel joint ventures in such fields as paints and solvents in Europe, and lengthy patent-exchange negotiations that culminated with the comprehensive alliance with I.C.I. in 1928–29. To some extent Du Pont was rebuilding the system of agreements that had prevailed in explosives in the prewar period, but the range and diversity of ventures represented a new dimension for Du Pont. The company continued, as always, to emphasize domestic markets, but there was a growing interest in the prospects beyond the borders, and a growing awareness of the potential impact of technological development and consolidation of the chemical industry in Europe.

Twenty years later, antitrust lawyers would represent Du Pont's foreign connections as part of a world-wide conspiracy to control international markets and prices and strangle would-be competitors at home and abroad. To support these charges, the government reviewed the entire history of Du Pont's foreign agreements going back to 1897 and introduced as evidence the observations of outspoken advocates of cartels such as Harry McGowan of I.C.I. on the purpose of these agreements. Documents subpoenaed from Du Pont and I.C.I. were also presented to demonstrate the accuracy of McGowan's comment in 1923 that the patent exchange provisions of these agreements were "a 'camouflage' to cover all relationships between the companies. . . ."[1] Yet Du

Pont's response to these charges, while not accepted by the federal courts in 1952, was not without some merit. During the 1920s at any rate, Du Pont's major aim in its negotiations with foreign firms was access to their technology. Implicit stabilization of markets was not, of course, objectionable to Du Pont as long as the arrangements did not directly contravene the antitrust laws; but these foreign negotiations must be seen in the perspective of Du Pont's general strategy in this period. The company had achieved success through diversification during the war and in the expansive mood of the postwar decade it continued to pursue that course. Later during the depression years the market stabilization aspect of these arrangements became more important. In this era of shrinking markets and widespread unemployment, governments in the United States and elsewhere shared the view of the big corporations that cooperation was necessary to stem economic decline. It is important, however, to distinguish the considerations that prevailed in this decade from those of the buoyant 1920s when Du Pont, and Americans generally, saw the prospects of "better living" through technological growth as virtually unlimited.

The primacy of the search for new technology is most apparent in Du Pont's relations with I. G. Farben. The Americans knew that they were behind the Germans in crucial technical areas and, like the British, could not view with equanimity the prospect of a German invasion of their home markets. But Du Pont was not prepared to concede access to or control over new technology simply in order to protect its market. In 1927 as negotiations with I. G. Farben were in a particularly (but not abnormally) acrimonious stage, Lammot du Pont defiantly asserted: "We neither fear your competition in this country nor doubt our own ability to stand up in the fight," adding "we would be better off in the end without it and are perfectly confident that you also would be better off without the fight."[2] The objective in this situation, as in many of Du Pont's contacts with I. G. Farben, was cooperation in technical development.

Du Pont's role in the establishment of a protective tariff for chemicals in 1919–22 must be seen in the same context. Du Pont lobbyists, including Irenee and cousin Coleman, now a U.S. senator, labored hard to push this bill through Congress, arousing critics who maintained that Du Pont wanted to establish a monopoly in organic chemicals, a charge repeated by hostile congressional investigators in the 1930s. The monopoly charge had little foundation since Du Pont shared the American organic chemical market with other companies, including Allied Chemical and Grasselli Dyestuff, which had passed back into German hands by the end of the decade. From Du Pont's point of view, the tariff served a tactical purpose, forcing foreign firms with technology that it hoped to acquire to deal with Du Pont.

Early Contacts, 1918–22

Du Pont had begun its move into organic chemicals in 1916 with its agreement with Levinstein in Britain and the construction of a big dye works at Deepwater, New Jersey. Lammot du Pont was put in charge of an Organic Chemicals Department to work closely with the Development Department in the new field. As early as 1918, however, the new venture was encountering technical difficulties and the Levinstein arrangement was generating more problems on the commercial side than was deemed worthwhile. At this point a new opportunity appeared.

When the U.S. entered the war, Congress passed a Trading with the Enemy Act that empowered the president to establish an "alien property custodian" to take charge of enemy assets that were considered potentially valuable to the war effort. Early in 1918 the Alien Property Custodian, A. Mitchell Palmer, proposed that his office be authorized not only to seize enemy property but also to dispose of it to ensure that it did not return to "enemy" control after the war. Amid the passions of the war, this novel assault on private property rights was accepted. The German chemical firms in the United States were the principal targets of Palmer's "Americanization" plans; but government officials discovered that the major assets of Germans in this area, aside from Bayer's Syracuse plant, were the patents to dyes and drugs that they had taken out to forestall American competition. Custody of these patents, however, had been given to the Federal Trade Commission. In November, 1918, as the war ended, Palmer again waved the bloody shirt before Congress, warning that the patents were the Germans' "secret weapon" through which they schemed to recover control of the American market. Once again Congress responded by passing an amendment to block the return of patents to German ownership.

Throughout this period Palmer's demands had been enthusiastically endorsed by the American Dye Institute, the lobby for the emerging organic chemical industry. In December, 1918, however, the A.D.I. began to have second thoughts about the course of events. The Bayer plant and patents were put up for public sale and acquired by Sterling Products Inc., a drug company, for $5.3 million. Sterling, uninterested in the dye patents, turned these over to its co-bidder, Grasselli Chemical Co. which had decided to enter the dye field. The entry of this outsider alarmed the big dye makers, National Aniline and Du Pont, both with heavy investments in their new ventures. A smaller rival might outflank them by picking up the other patents, might even make a deal with the Germans—as Grasselli was later to do.

In January, 1919, A.D.I. representatives, including Du Pont's dye specialists Poucher and Cesare Protto (another former B.A.S.F. sales agent) and John Laffey, the company's chief counsel, met with Palmer's lieutenant, Francis Garvan, who became custodian two months later when Palmer was elevated to attorney general; Garvan had in fact been the main architect of the custodian's patent policy and a supporter of "Americanization" of the dye industry. Together, the custodian's office and the A.D.I. arranged to establish a nonprofit organization, the Chemical Foundation Inc., that would become the repository of the dye patents held by the custodian and license them on a nonexclusive basis to manufacturers considered by the trustees of the Chemical Foundation to be untainted by German influence. Shortly after the establishment of the Foundation with financing provided by the A.D.I., the custodian began transferring to it some 5,700 patents for which the Foundation paid $271,000.[3]

Questions about the legality of the Chemical Foundation's actions appeared during the debates over the dyestuff provision of the Fordney tariff of 1922, but a full-fledged attack on the arrangement followed in 1923. The U.S. Justice Department sued the Foundation for return of the patents, arguing that they had been conveyed to the Foundation at a fraction of their true value, that the executive order authorizing the transaction was illegal, and the entire affair had been carried out for the benefit of a few large corporations whose precise relationship with the Alien Property Custodian raised questions about conflict of interest.

The origins of the suit are not entirely clear. Partisan politics probably played a role: as in the Muscle Shoals controversy, the Republicans were interested in publicizing suspected wrongdoings by the preceding administration. Irenee du Pont believed the main element was personal animosity between Garvan and the new Attorney General Harry Daugherty, who was later to face charges of improper involvement in the return of other confiscated properties to former German owners. At any rate, Irenee was sufficiently concerned about the case that he went to Washington to urge Daugherty and President Harding to drop it. His main interest was in suppressing charges that "the chemical industry . . . conspir[ed] to deceive the U.S. government." As an alternative he proposed that a congressional inquiry establish the true value of the patents.

Irenee's efforts to block the lawsuit were unsuccessful, but the trial proved to be less damaging than he feared. Federal district judge Morris who heard the case in Wilmington in the summer of 1923 dismissed the suit on all counts, observing in passing that, while the government had raised the issue of conspiracy on the part of chemical manufacturers to acquire the patents, it made no effort to support the charges and virtually ignored them in closing arguments. The government appealed the case to the circuit court and U.S. Supreme Court, but lost on both appeals. The Chemical Foundation and its industrial supporters emerged from the ordeal unscathed.[4]

The claim that Du Pont, through the Chemical Foundation scheme, "acquired the secret patents that became the key to its monopoly on dyes"[5] has become a staple element in popular mythology about the company. The Chemical Foundation patents did not in fact confer a monopoly on Du Pont since they were licensed on a nonexclusive basis and only about 10 percent of the chemical patents were licensed between 1919 and 1922. Furthermore, the German patents alone proved to be of little value since they required processing by specialists in coal-tar chemicals in order to produce uniform results: "Dyes would be synthesized in the laboratory, carried successfully through small scale experimental manufacturers, and then placed in production. One day the run of materials through the plant would produce a perfect product. Next day, using supposedly identical raw materials, formula, equipment and operators, the resulting dye might be off in shade or wretchedly low on yield obtained, for reasons then obscure."[6]

Du Pont quickly realized that it would have to go abroad in order to procure the "know-how" essential to work the dye patents. In spring, 1919, the company sent Leonard A. Yerkes and representatives of the Development Department to Europe to explore the possibilities of acquiring the know-how either by agreement with foreign companies or by recruiting trained chemists.[7] At this same time, Du Pont acquired the services of Eysten Berg, who had carried out a study of the synthetic nitrogen industry in Germany for the U.S. government at the end of the war and was well informed on technical developments in European chemicals in general. Berg also had many contacts with the I.G. and became an important liaison between Du Pont and the Germans in the ensuing months. In October, 1919, Berg informed Du Pont that Bosch of B.A.S.F. was interested in finding an American partner to establish a synthetic ammonia plant in the United States based on the Haber patents, and improvements made at Oppau during the war.[8]

Du Pont responded with interest, although it is unclear whether the Wilmington company was seriously contemplating an entry into the synthetic ammonia field at this point, or saw an agreement on nitrogen as paving the way for a more immediately useful arrangement on dye technology. When Du Pont sent Charles Meade to review the proposal with Bosch, the German pointed out that he could not discuss a dye agreement without the concurrence of other members of the I.G., but Meade was confident that "a successful ammonia arrangement certainly would lead at once to a dye exchange."[9]

Du Pont had approached B.A.S.F. at an opportune moment. The Germans were dismayed over the French demands to share ammonia technology or submit to the dismantling of the industry. Bosch had also been surprised and impressed by the action of the U.S. Alien Property Custodian in the sale of the dye patents. In November, 1919, Bosch met with Meade and Berg at Zurich and agreed to establish a joint venture with the Americans in nitrogen that would be a "world company" with restrictions only on the French, German, and possibly the British markets. Du Pont would put up the capital and hold 75 percent of the shares; B.A.S.F. would supply the technology and the specialists to design plants. Du Pont could congratulate itself on having outflanked the British, the only other chemical manufacturer with the resources to mount an undertaking of this magnitude. When Bosch met with Pollitt of Brunner-Mond shortly thereafter, he said he had already made a deal with the Americans.[10]

Good relations between the two companies rapidly soured, however, in the ensuing months. American sales agents of the I.G., such as Herman Metz, turned up in Germany and expounded at length on the role of Du Pont men, particularly Poucher, in the establishment of the Chemical Foundation and in lobbying for a permanent protective tariff on dye imports. Bosch was also angered when his application for a visa to enter the United States was denied, ostensibly at Poucher's behest, and by Du Pont's discussions with Brunner-Mond that seemed to undercut the Zurich agreement. By the summer of 1920 the projected world company for nitrogen was in ruins.[11]

Du Pont now turned to its alternative strategy, recruiting foreign chemists to man its dye works. The Yerkes mission in 1919 had already identified several potential candidates. During the winter of 1920–21 Du Pont hired four chemists from Bayer, and "spirited" them out of Germany together with "a truck containing drawings, formulas, and other important industrial information." The German government protested and sought, in vain, to block their entry to the United States. By the summer they were working for Du Pont, which announced an expansion of its dyestuff plant at Carneys Point.[12]

Berg resigned after this episode, and Du Pont's contact with the Germans virtually ended for several years although desultory talks continued. Before leaving Du Pont's employ, however, Berg had investigated an ammonia process developed by Georges Claude that was different from the Haber-Bosch method. Berg did not feel that the Claude process had been sufficiently tested on a commercial basis, but Du Pont kept the alternative in mind, and in 1924 established a joint venture with the French firm, L'Air Liquide, that developed the Claude process in the United States. As usual, Du Pont held a 63 percent majority of the shares in the venture, designated Lazote, Inc.[13] Du Pont continued, however, to be interested in the Haber-Bosch process, a consideration that contributed to later negotiations with I.C.I., and a resumption of relations with I. G. Farben in the mid-1920s. By this time, however, Du Pont's position in dyestuffs and organic chemicals was far stronger, as a result of its acquisition of technical know-how and the market stability established under the dyestuff provision of the Fordney-McCumber Tariff of 1922.

The Chemical Tariff, 1919–27

The protective tariff on dyestuffs instituted in 1916 and the wartime embargo on all German imports had stimulated the development of a domestic organic chemical industry that by 1919 was producing 55 million pounds of dyes, equaling the total volume of domestic production and imports in 1914. The report of the U.S. Tariff Commission in 1919, however, noted that American producers had been unable to develop on a large scale the "vat dyes" based on anthracene that were used extensively in the textile industry, and had not as yet achieved the 60 percent supply level stipulated in the 1916 act. Furthermore, after the armistice of November, 1918, German dye makers had begun talks with the War Trade Board of the U.S. State department, which had been authorized to license foreign chemical imports, proposing to reenter the American market to supply the vat dyes. Although the War Trade Board had established a committee to review this proposal that included such strong advocates of protectionism as Poucher and Dr. Charles Herty, American dyestuff producers feared that their wartime achievement would soon be undermined by a flood of foreign dyes once the embargo and high duties were lifted.[14]

The American Dye Institute commenced lobbying for a permanent tariff law that would set prohibitive rates on imports and extend the embargo for at least some time after the peace treaty was completed. In May, 1919, Representative Nicholas Longworth (Republican, Ohio) introduced a bill along these lines that had been drafted by Grinnell Jones of the Tariff Commission, who was sympathetic to the American dye makers, and representatives of the A.D.I. and the Chemical Foundation. In addition to imposing duties on dyes and intermediates ranging from 35 to 50 percent *ad valorem,* the bill would impose a per-pound surcharge, and would perpetuate the import licensing system administered by the War Trade Board.

The Longworth bill encountered opposition from the American Textile Institute, which was particularly unhappy with the licensing provision. Southern Democrats, who traditionally opposed protective tariffs, endorsed these criticisms; but more significantly, the Republican party, then in control of Congress, was split on the issue. Progressive-wing members from the Western states were suspicious of the links between the dye industry and "big businessmen" like the Du Ponts and Eugene Meyer. Other Republican leaders, including the influential Senator Boies Penrose and Philander Knox of Pennsylvania, seemed intent on smothering the Longworth bill so that they could advance the tariff demands of their industrial constituents ahead of the dye makers. The A.D.I. was able to lobby the bill through the House by September, 1919, but in the Senate it was stalled by these powerful opponents.[15]

Irenee du Pont entered the fray at this point, persuading Senator Knox to alter his views by arranging to have the licensing system transferred to the Tariff Commission; Knox thereafter became a strong supporter of dye protection. Other opponents proved less amenable, however, and the bill was held up by Penrose's Senate Finance Committee until the summer of 1920, at which point it was subject to a filibuster and expired along with the session. The only positive achievement of the A.D.I. at this point was the extension of the embargo to July, 1921.[16]

After this debacle, the A.D.I. regrouped to contemplate future strategy. Du Pont was now more actively involved in the effort as its hopes for an agreement with the Germans faded. In December, 1920, Congress reassembled in a special session to deal with the problems of the postwar recession, among other matters. The A.D.I. began immediately to push for an "emergency" tariff act to prolong the embargo and institute import rates at the levels proposed in the Longworth bill on a short-term basis, while also drafting a bill that would install a permanent system of protection. Du Pont's publicity bureau contributed to the effort with a battery of press releases emphasizing the military importance of dyestuffs and the actions of other countries, including Britain, France, and Japan to protect their organic chemical industries. By the end of May, 1921, the emergency bill had swept through Congress; the success was due in part to the fact that Western Republican members of the "farm bloc" had been forced to make concessions on industrial tariffs in order to get higher rates placed on agricultural imports in an effort to stem the rapid decline in farm commodity prices.[17]

The 1921 act provided a respite, but American dye makers still felt the need for long-term security. Charles Meade informed Irenee du Pont during the summer that the Treasury Department was under growing pressure to expand its import licensing which "will be utilized by the Germans to the fullest extent for attack in certain points."[18] The A.D.I. now conceived a new device to enhance the protective measures contemplated in the permanent tariff revisions under consideration in Congress. In the provisions of the Fordney-McCumber tariff bill relating to organic chemicals, the valuation of imports was to be made on the basis of the "American selling price" of the product, that is, the price at which it was currently being sold on the American market rather than the conventional valuation based on the price of the good at its point of origin. This measure undercut the cost advantage of foreign manufacturers, and reinforced the exceptionally high rates imposed: 60 percent *ad valorem* on finished products, and 55 percent on intermediates (to be reduced to 45 and 40 percent respectively by 1924) plus a 7-cent per pound surcharge.[19]

Over the next forty years tariff rates would fluctuate, moving progressively downward after 1945. By the mid-1950s the rates for intermediates averaged 21 percent and products based on organic chemicals fell in the 10 to 20 percent range; but the "American selling price" valuation remained in place, to the great irritation of America's trading partners in Europe. For American chemical manufacturers, however, the provision had become an article of faith. The A. D. Little consultant company estimated in 1962 that the American selling price restriction had held organic chemicals imports down to 20 percent of what they might have been otherwise, and had not seriously affected U.S. exports of chemicals after World War II. Free trade advocates in the 1922 debates over the Fordney bill and thereafter regarded the provision with loathing and blocked its extension to other products, but to Du Pont and other American chemical manufacturers it was an essential element not only in the development of the domestic market but also in defining their relations with foreign firms.[20]

The contest over the tariff had other less pleasant after-effects for Du Pont. As the bill proceeded, opponents focused increasingly on the "arrogance" of the dye lobby, and a special Senate subcommittee was set up to investigate the connections between the A.D.I., Du Pont, and the Chemical Foundation which was called a "subsidiary" of the Wilmington company. The investigation dissolved at the end of the 1922 session and the subcommittee never issued a report, but the charges lingered on, resurfacing during the Chemical Foundation lawsuit two years later, and were resurrected again during the Senate investigation of the munitions industry in 1934–35. Du Pont had been successful in achieving its legislative aims, but in the process had contributed to a public image that would create difficulties for the company in the future.[21]

Having secured protection and access to technical know-how in organic chemicals, Du Pont was content to develop its domestic position, acquiring a 32 percent market share in the U.S. field by the end of the decade. During the Ruhr crisis of 1923 Irenee had half-seriously suggested recruiting Bosch and the other leading I.G. chemists for Du Pont, but of course nothing came of this idea. Sporadic talks continued with the Germans who now were more amenable to a joint venture in dyes, but Du Pont remained cool. In 1924 Pickard wrote to Irenee that "while they [the I.G.] are making up their minds we shall be getting stronger and . . . as some of the small American companies are close to the rocks, our percentage of the American tonnage, aside from indigo, should be larger a year hence. . . ."[22]

The consolidation of I. G. Farben in 1925–26 provided an opportunity for reconciliation. During 1925, Du Pont and Nobel arranged to buy an equal block of shares in two German explosives firms D.A.G. and Köln-Rottweiler that had been part of the Nobel group before the war and were now desperate for new financing. The investment gave Du Pont and Nobel about 8 percent of the two German firms. In 1926 I. G. Farben acquired majority control of the two companies and absorbed them into its empire, so that Du Pont and Nobel now found themselves with shares in the I.G., although their combined holdings came to less than 1 percent of the German chemical firm. Since Du Pont and Nobel were also linked to the German explosives companies through the 1921 agreement on patent exchanges, however, the prospects for wider discussions were present.[23]

At this same time, Du Pont learned that Herman Metz, the former U.S. sales agent for Hoechst, had proposed a consolidation of all the German chemical importing agencies into a single firm, General Dyestuffs, Inc., which would also take over the dye business of Grasselli that controlled the former Bayer plant and patents. Metz appears to have initiated the merger on his own, to forestall a similar move by one of the other import agencies, but since this development came hard on the heels of the formation of I. G. Farben, Du Pont suspected that the German firm was directly involved. Irenee du Pont called on the U.S. Commerce secretary Herbert Hoover to urge him to take steps to block the merger by persuading American banks to deny I. G. Farben a $35 million loan. Hoover was sympathetic but could do nothing: investment in German business had become quite popular in the American financial community after the stabilization of the German monetary system under the Dawes Plan.[24]

By this time, Du Pont was committed to moving into the synthetic nitrogen field. In addition to developing the Claude process in 1925 Du Pont began buying into smaller firms in the U.S. including National Ammonia and the Michigan Ammonia Works, and acquired a rival European process, developed by Luigi Casale, that was believed to be superior to the Claude process, as part of a settlement of a patent infringement suit against the Niagara Ammonia Co., a subsidiary of a Swiss firm and the electrical utilities giant, Electric Bond and Share. Du Pont wanted both technology and distribution outlets; as Irenee later explained: "we wanted to collect all the customers we could."[25]

The move into ammonia was poorly timed. In 1926–27 agricultural prices began to slide into depression, and the bottom dropped out of the fertilizer market: prices for synthetic ammonia fell by almost 50 percent in that year, aggravated by the excess capacity developed in the immediate post war period.

All the big producers were hurt, not least the Germans. Both Du Pont and the I.G. could see the potential benefits to be gained from pooling resources and rationalizing operations in the ammonia and dye fields. Serious discussions were resumed early in 1927 that looked toward joint ventures in both these markets.

But by the end of the year the ammonia proposal had collapsed; negotiations continued on a joint dye venture, amalgamating General Dyestuffs and Du Pont's organic chemical division, but by 1929 this arrangement too had foundered. The main issue was control: the Germans wanted a 50 percent share while Du Pont insisted on a majority position similar to its arrangements with the French in cellophane and rayon. Since Du Pont was in the process of buying out its partners in these ventures, and since I. G. Farben would be providing Du Pont with access to the Haber-Bosch process and its organic chemical technology, the Germans saw little long-term benefit for themselves.[26]

The breakdown in discussions did not produce a rift as wide as the one that had prevailed in 1921–25. Du Pont and I. G. Farben agreed informally to respect each other's position in the dyestuff markets, and other American dye makers, including Allied Chemical, made similar arrangements that functioned through a system of mutual supplies of dyes and intermediates between the American companies and General Dyestuffs.

The Germans now proceeded to expand and consolidate their direct investment in the United States chemical market. When Grasselli Chemical Co. sold out to Du Pont, its dye business was taken over by a holding company, American I.G. Chemical Co. that was a subsidiary of I. G. Chemie in Switzerland, a company set up by I. G. Farben in that year to provide it with a refuge from future economic disruption in Germany, a farsighted move in retrospect. American I.G. also took over Agfa Ansco, which produced photographic film and assorted other properties, including a 50 percent interest in Winthrop Chemical. The former Grasselli dye works, renamed General Aniline, supplied General Dyestuffs, thus giving the Germans a foothold in the United States, bypassing the tariffs. By 1939 General Dyestuffs had a 21 percent share of the market, exceeded only by Du Pont and Allied Chemical in volume, and actually leading them in terms of value of products sold. The "Chinese wall" of the 1922 tariff had thus been effectively breached.[27]

After this second impasse with the Germans, Du Pont turned to its traditional allies in Britain. Imperial Chemicals was not interested in entering the American dye market and had, among other things, the Haber-Bosch patents in ammonia. Moreover, the British firm was eager to construct an edifice of agreements on the existing foundation of patent agreements in explosives and joint ventures in Canada and South America. In 1928 Du Pont sent Fin Sparre to London to begin talks at the technical level that would lead to the "grand alliance" which would be the central feature of Du Pont's foreign relations over the next two decades.

CHAPTER NINE

Du Pont's Foreign Relations: The "Grand Alliance," 1919–29

Although the German chemical companies maintained an undisputed technological edge over the rest of the international chemical industry, the other European producers also had much to offer the Americans in the immediate postwar years. Du Pont's technological insecurity resulted in a shuttling of corporate executives between Wilmington and the European industrial centers throughout the 1920s. The objective of these sometimes frenetic American forays to London, Frankfort, and Paris was a series of patent and process agreements which would unlock the secrets of foreign manufacture. The Yerkes mission to Paris during the winter following the Armistice was merely the first of many such trips abroad by Du Pont representatives.

Although Du Pont was particularly interested in gaining access to the technology associated with dyes and synthetic ammonia during the early months after the war, the director of the European mission was fully cognizant of the Development Department's long range plans for overall expansion beyond explosives. Once in Paris members of this group began to urge Wilmington to establish more permanent representation there to study European chemical developments. Yerkes urged the establishment of "a real office in Paris" since "I believe it is infinitely more important to know at first hand and promptly what is going on in Europe than it is for instance for us to know what the National Aniline Company are doing in this country." With the cosmopolitan Irenee taking charge at Du Pont this advice met a generally favorable reception.[1]

Paris proved to be a wise choice for an expanded Du Pont office, for during the immediate postwar years negotiations for foreign patents and processes were most successful with the French. Yerkes's 1919 mission was planned to gain additional information on a number of fields, including cellulose—a natural offshoot of Du Pont's concentration in explosives. Using cellulose as a chemical base, manufacturers could produce a wide variety of products just beginning to reach consumers.

Du Pont explored a number of these developments with the major French producers, but its principal achievement was an agreement in January, 1920, with the French textile conglomerate, the Comptoir des Textiles Artificiels. This agreement gave Du Pont access to the viscose process used in the production of artificial silk. Armed with this information Du Pont established a subsidiary, Du Pont Fibersilk Company (later renamed Du Pont Rayon Company), to exploit the patents in the United States. Yerkes, the company's chief negotiator with the Comptoir, was selected to head the new venture. The French were more than willing to share their technology, since rayon had great potential in the American market. Du Pont would build the plant, manage it, and establish a marketing and distribution program. The French received a 40 percent share of the stock of the venture, 24 percent for their know-how and purchasing an additional 16 percent. Within a decade the French investment in America proved its worth. American consumption of rayon increased from 8.2 million pounds in 1919 (most of it produced by American Viscose, a subsidiary of the British firm, Courtaulds, that had begun operations in 1911) to over 130 million pounds by 1929. Most importantly, Du Pont's entry into rayon production was a milestone in the company's diversification. Rayon manufacture consumed an enormous amount of chemicals produced by Du Pont, including sulfuric acid, caustic soda, carbon bisulfide, acetic acid, and acetone.[2]

Moreover, Du Pont's successful negotiations with the Comptoir paved the way for an agreement on another miracle product of the postwar years: cellophane. The Swiss chemist Jacques Brandenberger had developed cellophane in 1912, but it was not until the end of World War I that he was able to obtain the necessary financial backing from the Comptoir. Soon after the rayon agreement was signed, French interests organized La Cellophane to exploit Brandenberger's product. After months of bargaining W. C. Spruance of Du Pont and Charles Gillet from the Comptoir worked out arrangements to form a Du Pont subsidiary to take over La Cellophane's promising North American business. As it turned out, the French patents on cellophane were much weaker than French claims in the rayon field, a situation that Du Pont soon exploited.

Within three years after the agreement with La Cellophane was signed, Du Pont challenged the value of the patents. The dispute eventually led to Du Pont's purchase of the minority French interest in 1929.

Cellophane was a major commercial success from its earliest years and a substantial contributor to Du Pont's earnings through the 1950s, second only to nylon. Within two years after opening the Buffalo, New York, plant, Du Pont was bringing in a return on investment in excess of 50 percent. To head off potential competition from Sylvania Corporation, a U.S. subsidiary of a Belgian firm that manufactured its own variant of transparent packaging, Du Pont in 1930–31 developed a moisture-proof cellophane and began promoting its use for such novel products as fabrics and draperies. In 1933 Du Pont sued Sylvania for patent infringement, settling out of court when the rival firm agreed to share future technical developments with Du Pont in exchange for an exclusive license to sell the moisture-proof material, an arrangement that eventually led to an antitrust suit against the two firms. Meanwhile, Du Pont retained its monopoly of cellophane in the United States by continuous technical improvements and reductions in manufacturing costs.[3]

The crowning achievement of Du Pont's business diplomacy with the French came in 1924 when it acquired the American rights to the Claude process for production of synthetic ammonia. When production of ammonia began at Du Pont's mammoth new plant at Belle, West Virginia, in February, 1926, the company held the key to eventual development of urea, methanol, ethylene glycol and other higher alcohols, nylon intermediates, and methyl methylacrite resin ("Lucite"). Ammonia could also be used in the production of rayon and nitric acid. Thus by 1926 Du Pont was no longer dependent on nitrate supplies from Chile. This productive capacity in high pressure synthesis was enhanced in 1927 by Du Pont's acquisition of the American rights to the Casale process.[4]

Meanwhile, Du Pont resumed its prewar contacts with the British. Sir Harry McGowan's visit to the United States in the spring of 1919, for discussions connected with the renewal of the 1914 oral agreement on explosives between his company, Nobel, and Du Pont, led to the Patent and Process Agreement of 1920, whereby both companies arranged to share all new developments in explosives. By mid-decade that agreement proved to be the source of friction between the two parties, however, since the British insisted that in addition to being a patent agreement, the 1920 arrangement constituted a commercial deal to allocate territories for exports between the two firms. Du Pont executives adamantly denied this interpretation, pointing out that they had taken pains to construct the agreement within the legal confines of American antitrust laws.

H. G. Haskell, who headed Du Pont's Explosives Department in the postwar years, and had been privy to Du Pont negotiations with Nobel since 1907, maintained that Nobel's stance in 1923 reflected its concern over Du Pont sales of nitrocellulose powder in continental Europe. Haskell emphasized that Nobel did not even manufacture this particular powder, hence it was inconceivable that the agreement should apply. He added that the ten-year limit placed on the 1920 agreement was evidence that the parties recognized that industrial relations would necessarily change over time, making it impossible to establish rigid boundaries either in terms of market territories or sales products.[5]

Despite the disagreement over military powder, relations between the two explosives companies were decidedly friendly after the war. The 1920 agreement had provided for possible exchanges of information on products other than explosives, and Du Pont was hopeful that patient discussions would yield valuable knowledge in some of the lines being entered by the British. As always, synthetic ammonia topped Du Pont's list, but here the British were tight-lipped. Nevertheless, Haskell and other Du Pont observers were struck by the similarities between the diversification programs planned by the two companies in the early 1920s. By 1922 both Du Pont and Nobel were involved in dyestuffs, leather substitutes, auto accessories, film, varnishes, and acids as well as explosives.

The British, however, were unimpressed by Du Pont's offer to share patents and processes in the new fields at this point, since they were cultivating relations with the Germans. McGowan pursued a Du Pont connection for other reasons. Nobel was particularly anxious to avoid competition in its home market, the British Empire. Moreover, the British were eager to "rationalize" sales elsewhere, especially in Latin America, where the market for explosives, both commercial and military, remained promising. Nobel, with its more pronounced awareness of international export markets, realized that the political and economic nationalism emerging in that region would undoubtedly prompt desires for economic development. Nobel planned to be there to assist in the establishment of an indigenous chemical industry.

As early as 1917 Nobel began to press Du Pont for agreements on exports to South America and for joint ventures wherever the industrial outlook seemed to warrant local production. Du Pont was lukewarm about these proposals at first, but by the end of the war the Americans embraced the concept of organizing the Latin American market. Initially, Du Pont was motivated by a desire to continue its policy of cooperation with Nobel and other prospective rivals. Since South America promised to be a commercial battleground among the British, Americans, Germans, and others after the war, Du Pont accepted British overtures to regulate the arena. At the same time Du Pont entered the 1920 Patent and Process Agreement, it negotiated a "South American Pooling Arrangement" with Nobel. Both parties agreed to divide equally all profits from the sales of commercial explosives in South America except in Chile, and to exchange information on orders for military explosives from that region. No restriction, however, was placed on prices or volume of sales, in order to avoid the charge that this was a territorial agreement.[6]

In addition, the two companies set up a joint venture to manufacture and sell commercial explosives in Chile. To avoid future trouble in the Chilean market, Nobel and Du Pont invited the Atlas Powder Co. to join the new undertaking, called the Compania Sud Americana de Explosivos (C.S.A.E.). Du Pont and Nobel held 41½ percent each of the capital stock, with Atlas taking 15 percent, representing roughly the proportion of commercial explosives sold by the three companies in Chile and Bolivia. After 1923 the three parent companies restricted exports of explosives to those two countries to the extent that C.S.A.E. could manufacture for the local market more cheaply than any of the three parties. Thus by mid-decade Nobel and Du Pont had quietly settled their differences in one of the principal markets for mining explosives in the world.[7]

It was in Canada, however, that Nobel and Du Pont worked out their most lucrative joint venture. Their cooperation in the Dominion dated back to 1911 when an agreement had been made to handle manufacture and sale of commercial explosives in Canada and Newfoundland through Canadian Explosives Ltd. (C.X.L.). Use of C.X.L. as a conduit for the two parent companies extended in 1920 when a new agreement was drafted granting C.X.L. a non-exclusive license for patents and processes in specified explosives and chemical products in Canada and Newfoundland. The Canadian company was enormously profitable in this era of mineral exploitation in Canada. By the end of the decade Nobel regarded C.X.L. as the heart of its relationship with Du Pont, and installed at its head Arthur B. Purvis, a man with wide experience in international commercial negotiations who would later head the Allied Purchasing Commission in the United States during World War II.[8]

The Export Debacle

While Du Pont was weaving an international fabric of patent agreements and joint ventures, and protecting its home front by lobbying effectively for a tariff on chemical imports, the company devoted some attention to internal reorganization of its foreign relations. Initially, concern focused on the management of routine contacts with Nobel. As early as 1917 R. R. M. Carpenter complained to Pierre du Pont that cables between Nobel and Du Pont had "a hard time finding a resting place as they are sent from one Department to another until they find someone who knows something about it."[9] Carpenter's pleas for centralization of Nobel affairs within the company were partially resolved in 1919 when he was given authority to oversee all correspondence with the British regardless which department was involved. Within a year further progress was made on this score. Du Pont sent Jasper E. Crane, formerly with Arlington Company, to London to act as the company's European representative. Crane reorganized the London office to handle not only Du Pont's purchasing requirements but also overall surveillance of European chemical developments. An effective administrator and negotiator, Crane was soon praised as a "strong, live wire" in control of the situation.[10]

The move to centralize foreign relations was also a reflection of Du Pont's more aggressive pursuit of export markets after the war. The principal impetus for this move came from the desire to cut the inevitable losses foreseen at the end of hostilities. This time Du Pont was determined to be prepared for the down side of the boom and bust cycle in the explosives industry. In its search for markets in South America and East Asia Du Pont was following a well established pattern set by its European counterparts. For decades the British and German chemical firms had relied on exports as the safety valve enabling them to increase output without creating havoc in their primary markets.

The immediate postwar outlook was particularly promising. The Germans and British had been preoccupied by the war, enabling Du Pont to enter some of their traditional overseas territories for the first time. Moreover, sales to the Allies and the American government had spurred construction of new manufacturing facilities that gave the company an extra incentive to begin thinking in global terms. Du Pont was particularly encouraged by rising export sales figures as disruptions of international shipping routes delayed a revival of European exports.

In the spring of 1919 Du Pont organized a subsidiary to handle its export sales. This venture was largely the brainchild of F. W. Pickard, vice-president in charge of sales. Pickard had come up through the ranks via military sales, the one area that required a cosmopolitan outlook, given the company's active solicitation of export markets in military powders. Pickard was bullish about Du Pont's export possibilities and became the company's leading advocate of centralization to boost sales abroad, to parallel the large international sales organizations maintained by the European firms.[11]

Du Pont's most ambitious undertaking in exports involved the dye trade in China. The forced retreat of German firms from East Asia in World War I created a vacuum that Du Pont's dyestuff specialists, Meade and Poucher, regarded as an ideal outlet for surplus production from the Deepwater plant. Profitable sales of these "vat dyes" could be used to subsidize the company's diversification into the more sophisticated branches of organic chemistry. Within a year after the export company was established, however, the bottom fell out of the international dye market. By the fall of 1921 the export company was barely breaking even, and the attempt to push into the China market was costing the company over $2,000 a week.[12]

Beset by the recession that began in late 1920, Du Pont's enthusiasm for overseas adventures quickly ebbed away. By 1922 the export company had been dismantled and Du Pont returned to its decentralized system for handling export sales. There were several reasons for this failure beyond the immediate economic slump. Although the postwar recession hit all of the chemical industry, it paved the way for a speedy recovery by the Germans of their prewar position. When recovery began in 1923 the German firms had reestablished their trade network, particularly in China. Moreover, with the exception of Pickard, Du Pont's top management had traditionally been cautious about expansion outside the United States. Pierre's reluctance to move abroad in 1907 was passed on to his associates who tended to associate Du Pont's strength with its position in the home market and feared the capabilities of the Europeans. Irenee du Pont, upon learning of the export company's precarious financial position in 1921, observed to the head of the Dyestuff Department: "If we cannot compete with Germany in this country without an embargo or a duty which spells embargo, how can we possibly compete with them in China?"[13]

Finally, despite Pickard's attempt to centralize Du Pont's export sales after the war, the company remained committed to a corporate structure which dispersed authority along product lines, particularly after the 1921 reorganization. Given this predisposition, it was natural for general managers of the various departments to conduct their own market surveys and overseas sales efforts. Even during the early stages of the export venture, exports remained an appendage to the company's domestic business. Until the 1950s Du Pont exports never comprised more than 6 percent of the company's total sales. During the interwar years British and German overseas sales dwarfed Du Pont's export figures, constituting between one third and one half of their business in the 1930s.[14]

By mid-decade, therefore, Du Pont abandoned its short-lived attempt to chart an independent course abroad, and returned to its traditional foreign policy. Whenever a foreign market looked particularly alluring, Du Pont turned to arrangements with other firms which would enable it to manufacture its products abroad either by means of joint ventures or by cross-licensing agreements.

By the end of the 1920s, Du Pont's European strategy to acquire patents and processes had achieved the desired effect. The company's synthetic ammonia plant at Belle helped to boost national production of ammonia products from 113,700 net tons in 1923 to 172,000 net tons in 1927. Rayon and cellophane production at the Buffalo plant was likewise well under way. The company's research and development effort had also begun to bear fruit, holding the promise of additional leverage in bargaining with its European counterparts. Duco was only the first and most highly acclaimed Du Pont development; notable improvements in rayon, cellophane, and other new products were also being introduced. Du Pont realized early in the 1920s that its strongest suit in foreign negotiations was its technological and market strength at home.

Expansion of the Anglo-American Entente, 1926–29

The consolidation of the European chemical industry was the most significant development of the mid-1920s for Du Pont's foreign relations. The reorganization of I. G. Farben posed a formidable threat to Du Pont's carefully built position within the American market. The Americans were even more anxious about the I.G.'s potential for research advances now that consolidation had been achieved. By 1927–28, with the establishment of the American I.G. and stalemate in patent and process discussions with the Germans, Du Pont's apprehensions appeared to be well founded.

Meanwhile, however, relations with the British seemed to be yielding concrete results for both sides. A new agreement on explosives was completed in 1926, and the Canadian joint venture had proceeded so smoothly and profitably that a new contract providing for further cooperation was signed at the same time. Du Pont also offered Nobel a license to its Duco auto finish process, which led to the establishment of a new subsidiary, Nobel Chemical Finishes Ltd., that provided a model for subsequent Duco licensing arrangements with French, Italian, and German manufacturers.[15]

For a time in the middle of the 1920s Du Pont had pursued simultaneous negotiations with both of the new European chemical giants, I.G. and I.C.I. By the end of 1927, however, it had become clear to both English-speaking firms that I.G.'s confidence in its technical superiority and aggressive approach to foreign markets provided insurmountable obstacles to any comprehensive partnership arrangements. Du Pont and I.C.I. executives began to discuss the possibilities of broadening the scope of their earlier agreements to provide mutual protection in chemical fields where the I.G. challenge affected both companies.

The proliferation of agreements with foreign firms and prospects of even greater cooperation with I.C.I. led to a resurgence of interest among Du Pont executives for further coordination of foreign affairs within the company. The progression of consolidations of national chemical industries throughout Europe prompted the realization that "policies adopted by one industrial department of the Du Pont company very often affected the interests of other departments."[16] The decentralized administration of industrial affairs that was a corporate strength at home could easily become a liability abroad. It was conceivable that the technology Du Pont had received from the French, for example, in the field of ammonia synthesis or rayon could mistakenly flow to I.C.I. as a result of an ill-phrased patent clause. The company also suffered embarrassment when its left hand did not know what the right hand was doing. One incident that especially rankled Du Pont's top echelon was an occasion in which the company's Sales and Purchasing departments, working independently, promised patent rights to Duco to two different Italian firms.[17]

In 1927 a subcommittee of the Executive Committee was set up as an advisory body on foreign policy. General managers and heads of subsidiary companies were instructed to notify this committee prior to taking any action that involved Du Pont's foreign affairs, to prevent misunderstandings and confusion resulting from the intricacies of the company's technological exchange agreements. The committee included Pickard, Crane, and H. A. Haskell, all of whom had wide experience in foreign fields.

The first task of the new committee was to prepare for a series of meetings in December, 1927, between Du Pont and I.C.I. in London. At this conference the prospects for a full exchange of information in fields outside explosives was raised. At this time Sir Alfred Mond demurred about a comprehensive agreement with Du Pont, fearing that such an arrangement might jeopardize I.C.I.'s attempts to gain access to dyestuffs and nitrogen technology from I.G., or that it would imperil its alliance with Allied Chemical in the alkali field in the United States. Six months later, however, Mond was more receptive to a partnership with Du Pont. By then relations with I.G. had soured sufficiently to jar the entire I.C.I. establishment into the realization that any general agreement with the Germans would require the British to relinquish control over major sectors of the nation's chemical industry. Of the two possible American alternatives, Allied Chemical or Du Pont, the latter had pursued a more ambitious diversification strategy, thereby offering I.C.I. a more promising list of chemical processes. By October, 1928, Mond, now elevated to Lord Melchett, had agreed with McGowan to sell I.C.I.'s interest in Allied, thus removing a last major obstacle to the extension of the 1926 patent agreement with Du Pont.[18]

Throughout the spring of 1929 Du Pont's foreign relations committee, together with the Legal and Development departments, worked out a basic agreement with I.C.I. representatives. This document, called the "London draft," provided for a clear delineation of markets, with Du Pont's exclusive territory North America (excluding Canada and Newfoundland which belonged to C.I.L.), and I.C.I.'s exclusive territory the British Empire (again excluding Canada and Newfoundland). The rest of the world would be "non-exclusive," in which both partners were free to exploit all products covered by the agreement. Since the London draft seemed to comprise more of a commercial agreement than a patent exchange, Du Pont lawyers labored over it to ensure that the wording conformed to American antitrust laws. In November, 1929, the final draft was signed.

The list of products covered by this agreement was indeed comprehensive, including almost all products then being produced by either company: "non-military explosives, cellulose derivatives (plastics, lacquers, and film), paints and varnishes, pigments and colors, acids, fertilizers, synthetic ammonia, synthetic products of the hydrogenation of coal and oil, dyestuffs and other organic chemicals, alcohols, insecticides, fungicides, and disinfectants."[19] Du Pont's agreements with French firms prevented it from including cellophane and rayon in the I.C.I. agreement. Similarly, I.C.I.'s long-standing relationship with Allied Chemical mandated the exclusion of alkalis from the list. Both companies also excluded the entire field of military explosives, since government regulations on both sides of the Atlantic made such an arrangement impossible. The shadow of I.G. was cast over the final document, since neither party wanted the new agreement to hamper their respective negotiations with the Germans. Thus the agreement explicitly provided for modification of the clauses relating to dyestuffs if either party could reach an agreement with I.G.

Fifteen years later Justice Department lawyers would focus on this agreement to demonstrate charges that Du Pont and I.C.I. had conspired to eliminate competition in world chemical markets. But the context in which the "Grand Alliance" of 1929 was established must be kept in mind. Although both parties hoped to establish an understanding within which their home markets would be protected and provisions would be made for an orderly exploitation of new chemical technologies, Du Pont took pains to make the agreement conform to American antitrust laws as they were understood in 1929. John K. Jenney, secretary of the Du Pont foreign relations committee at the time, maintained that: "It was the opinion of our lawyers that it was perfectly legal to relate commercial restrictions to patents. . . . It was legal to license a patent or a secret process on an exclusive basis, which had the effect of preventing the export by the grantor of the patent license of a product covered by that patent or secret process."[20]

The *form* of the agreement entered into by Du Pont was an accurate reflection of the *substance* of that agreement as perceived by its top management. Du Pont viewed the 1929 agreement primarily as an instrument for the exchange of valuable technical information. At the time the agreement was signed, Du Pont was especially anxious to gain access to I.C.I.'s synthetic ammonia technology, so much so that Jasper Crane persuaded I.C.I.'s Executive Committee to allow a Du Pont contingent into their Billingham plant even before Du Pont signed the document. I.C.I., too, was undeniably eager for Du Pont patents, but its priorities were more blurred between the concepts of exchanging technical information and sharing markets. I.C.I., untrammeled by national antimonopoly laws comparable to those in the United States, viewed the arrangement with the Americans much as they viewed arrangements with other large chemical manufacturers, which in the late 1920s consisted primarily of cartel agreements. The differences between the two companies in their interpretation of the agreement was evident in their debates over a compensation clause in the final draft. By this provision the value of each party's contribution of new technology would be periodically assessed. I.C.I. initially objected to this clause on grounds that in all likelihood each party would be contributing equally valuable patents and processes. Du Pont, however, stood firm, arguing that the absence of such a provision would make it appear that the agreement was a patent pool rather than a strict business contract. I.C.I. capitulated, and in 1930 Du Pont received the first payment after appraisal, amounting to $97,751.30. Another evaluation was conducted in 1936.[21]

The Patents and Processes Agreement of 1929 has been called "the lodestone for Du Pont's and I.C.I.'s 'foreign relations.' "[22] But the association between the two companies went far beyond this point, extending into a wide range of cooperative ventures all over the globe. During the middle of the decade, the Americans and British made provisions for the return of the Germans into the South American explosives field, providing their erstwhile partner Dynamit A.G. a 25 percent share of sales of C.S.A.E. in Chile and establishing a new joint subsidiary, Explosives Industries Ltd. (E.I.L.) to cover the rest of South America. Du Pont withdrew from E.I.L. in 1938 in order to avoid legal problems with the U.S. Justice Department over the neutrality and antitrust laws.[23]

The Anglo-American chemical entente also extended in East Asia. Following the demise of the Du Pont export company's attempts to establish a dyestuffs foothold in China, it had assumed a low profile there vis-à-vis European chemical exporters. By the late 1920s Du Pont had reached an agreement with I.C.I. that provided an exception to the territorial arrangements established in the 1929 Patent and Process Agreement. In order to give Du Pont a stronger negotiating position with I.G., I.C.I. permitted the Americans to retain sales markets in its traditional areas, China and India, until 1931.[24]

In Australia, a proposed tariff on imports of leather substitutes and rubber-coated fabrics, items manufactured by both I.C.I. and Du Pont, stimulated the two companies to establish a joint venture in 1926, called Nobel Chemical Finishes (Australasia) Ltd. Du Pont pulled out of this operation a decade later, and in the meantime had withdrawn as well from the synthetic ammonia field in Australia. In Canada, the profitable arrangement established through C.X.L. was extended into heavy chemicals and other products by the end of the 1920s, and under the 1929 Agreement the Canadian firm, rechristened Canadian Industries Ltd. (C.I.L.), was assigned licenses to virtually all of the patents and processes exchanged by Du Pont and I.C.I., with the provision that it would keep out of export markets.[25]

The success of these various ventures—in South America, the Far East, and especially in Canada—helped reinforce the links between the two companies forged by the 1929 general agreement. They also encouraged the partners to expand their overseas operations in the following decade when the depression eroded the home markets of both Du Pont and I.C.I. Through its I.C.I. connection, Du Pont was able to take tentative steps abroad without experiencing directly the hazards of market instability and unrestrained competition that had contributed to the collapse of its export venture directly after the war.

Final Steps in Foreign Relations, 1927–30

The late 1920s was an extraordinarily hectic time for Du Pont executives charged with managing the firm's foreign relations. In addition to the complicated negotiations undertaken with I.C.I. and I.G., Du Pont had to monitor patent agreements with other firms and attempts by foreign competitors to invade its home market. By 1929 these problems prompted an administrative reorganization in Wilmington to cope with the complexity, and a new strategy to meet the import threat.

The most serious challenge to Du Pont's American market existed in the fields of rayon and cellophane. By 1926, just two years after artificial silk had been given its new name, American producers were worried about a rayon glut in the United States. Although there were only two American manufacturers when Du Pont acquired the French viscose process in 1923, by the end of the decade there were over a dozen. A more serious problem was the steadily growing volume of imports, principally from Italy and Japan. The phenomenal rise in American demand was irresistable to foreign manufacturers, and pleas for further tariffs encountered greater resistance from Congress than earlier in the decade.[26]

Du Pont worked out a four-pronged strategy to overcome this foreign threat to its rayon market. First, production at the Buffalo plant was expanded to provide Du Pont with sufficient capacity to undersell prospective competitors. Second, Du Pont cultivated cooperative relations with foreign and domestic rayon producers, in particular the largest U.S. rayon manufacturer, American Viscose, with whom Du Pont had coordinated market research since the early 1920s. Third, by mid-decade Du Pont concluded that growth in the rayon field was most likely to occur in specialty lines using advanced technology. In 1927 Du Pont purchased the North American rights to manufacture rayon using the acetate process perfected by the French firm, Societe pour la fabrication de la soie Rhodiaseta, which was jointly owned by the Comptoir de la Textiles Artificiels and Rhone Poulenc, one of the largest French chemical firms. Du Pont further increased the range of rayon products it manufactured in 1928 by acquiring the American rights to "Celta," another permutation of the rayon process, from the Alsa S.A. of France.[27]

Du Pont's most novel, and ultimately most effective, response to market pressures in the rayon field was the promotion of internal research as a complement to business diplomacy. Artificial silk had been one of the first areas investigated by the Development Department in its diversification plans at the end of World War I, and Du Pont continued to push for improvements in this area, including a research program on the whole field of synthetic fabrics based on cellulose. By the middle of the 1920s Du Pont scientists had begun to move further, toward the development of an entirely synthetic fiber, not based on cotton or wood cellulose. Nylon, the major result of that research, would be the most important product to come from Du Pont in the following decade.[28]

Du Pont's spectacular growth as a diversified chemical company in this era was due in no small part to the successful execution of an intricate foreign policy. Unlike its European counterparts, however, Du Pont never fully centralized its administrative machinery primarily around international markets. Since the export fiasco in Asia Du Pont shied away from any departure from its proven strength in letting each product department carry full responsibility for exports. It was only in the acquisition of foreign technology and know-how that the need for centralization became urgent. Although unwilling to reorganize its foreign business along the lines of I.C.I., Du Pont in 1930 did establish a Foreign Relations Department to supplement and eventually take over the duties of the Foreign Relations Committee.

The 1929 Patents and Processes Agreement was the occasion for the formation of this new auxiliary organization. Although the Du Pont Foreign Relations Department never achieved influence in the company comparable to that of I.C.I.'s Foreign Trade Group, by the mid-1930s its duties ranged far beyond the British alliance. Under Wendell Swint, formerly an assistant to Jasper Crane in London, the Foreign Relations Department became an essential service unit controlling connections between foreign firms, synchronizing the operations of joint ventures abroad, and supervising Pickard's second effort to drum up substantial foreign trade through the establishment of a Foreign Trade Development Division in 1928. Although the Foreign Relations Department lacked the resources of the manufacturing departments, it survived the interwar years as a testament to Du Pont's continuing reliance on a healthy foreign policy to ensure prosperity at home.[29]

CHAPTER TEN
Du Pont in the Cartel Era

As the Great Depression struck industry after industry, nation after nation, businessmen sought refuge in ways that had historically provided some measure of relief. For national industrial groups this generally took the form of reinvigorated efforts to form trade associations to eliminate "cutthroat competition" in times of economic distress. In the United States the Roosevelt administration's flirtation with the National Recovery Act indicated a willingness to let business regulate itself by industrial codes of "fair competition." The failure of these measures to pull the economy out of the depression produced an increasingly strident outcry against big business in the United States, exacerbated by the government's shifting attitudes toward the nation's industrial leaders after 1934. By the end of the decade the conviction that responsibility for the nation's economic woes rested in large part on the actions of big business stimulated a revival of antimonopoly sentiment in the United States, a trend further aggravated by the onset of World War II. The early 1940s witnessed an unprecedented growth of government investigations to determine the causes of the nation's lack of preparedness for war. Among the suspects were the nation's largest corporations.

Industrialists in Europe also redoubled their cooperative efforts to cope with their difficulties in the depression. For European producers, however, there were more options. In addition to unification within each national group, they could also rely on international cartels to regulate trade on a much larger scale. Since membership in a cartel invariably involved participation in voluntary curtailment of production and the establishment of quotas and/or divisions of market territory, U.S. firms could not legally join such organizations. Nevertheless, many American companies found ways to cooperate with their international colleagues just shy of outright membership in cartels. Whether or not such discreet cooperation with foreign producers actually helped American industry to ride out the business cycles of the interwar years is an unresolved issue. But these relationships clearly contributed to the charges of complicity leveled against American corporations in the 1940s when questions of national security were paramount in the public mind. The international chemical industry was a central subject of these controversies and became a textbook case for studies of the economic and political complexities intrinsic to interwar business diplomacy.

The Du Pont–I.C.I. Alliance, 1929–44

The depression encouraged chemical manufacturers on both sides of the Atlantic to augment their already well-developed efforts at international cooperation. As always, Du Pont and I.C.I. courted I. G. Farben, with about the same desultory results they had experienced through the 1920s. The impetus for chemical diplomacy in the postwar decade had been chiefly to gain access to technology in the reasonably well-established fields of dyestuffs and ammonia. The 1930s, on the other hand, witnessed a veritable revolution in the chemical industry with developments in synthetic forms of rubber, plastics, and textiles appearing with startling rapidity. For the first time the Americans were approached by their European counterparts as much for their technological insights into these new areas as for their ability to help stabilize international markets. Embodying this spirit of transatlantic cooperation, Sir Harry McGowan wrote to Walter Carpenter, Jr., of Du Pont in 1931: "The day for isolation is over, and the more companies like yours and ours pool our knowledge, the greater the world will be benefitted."[1]

As the focal point for the Du Pont–I.C.I. alliance, the Patents and Processes Agreement of 1929 theoretically served as the conduit for an easy flow of technological information between the two parties. The chief benefits to Du Pont in the early years of the 1930s were in the coveted fields of synthetic ammonia and dyestuffs. Du Pont sent technical experts to England to observe I.C.I. operations, such as John Brill, who became chemical director of Du Pont's Ammonia Department after spending a year at the Billingham plant. For I.C.I., the early benefit of the Agreement consisted primarily of the confidence it gave them that the international chemical market would not be disrupted by Du Pont adventures in Europe or elsewhere. I.C.I., far more than Du Pont, cherished the increased number of joint ventures which the Agreement encouraged. In addition, the licensing provisions ensured that Du Pont would keep out of European markets that I.C.I. was seeking to develop.

Although both sides lauded the success of the 1929 agreement, their relationship was not all smooth sailing during the turbulent 1930s. People in both companies periodically grumbled about the unequal advantages accruing to the other partner under the arrangement to share technology. Within five years Du Pont managers were complaining about I.C.I.'s narrow construction of the exchange agreement. Moreover, relations grew more precarious as I.C.I.'s negotiations with the Germans yielded some concrete gains. Although uninterested in the synthetic fuel field, Du Pont was irritated by I.C.I.'s readiness to exclude it from that company's extensive dealings with I.G. in the development of the hydrogenation process. Wendell Swint of Du Pont's Foreign Relations Department echoed the dissatisfaction of those working at the departmental level of Du Pont when he complained to the I.C.I. representative in New York that "every time a subject now covered by the ICI/Du Pont (agreement) is withdrawn from that agreement, some of the force is taken away from it."[2]

By the end of the decade numerous other products had been discussed as possible candidates for cross-licensing arrangements under the 1929 agreement, including fluorine compounds, polymers, safety fuses, anhydrite plasters, and urea formaldehyde. As both firms moved into the "miracle products" of the 1930s, plastics and synthetic resins, each gained from the exchange of information. Du Pont was particularly pleased to have access to I.C.I. research in polyethylene and methyl methacrylate resin ("Lucite"). For their part, the British were among the first to receive licenses from Du Pont for the breakthroughs in nylon and neoprene (synthetic rubber).

By mid-decade developments in the chemical fields were occurring so quickly that both parties were sometimes confused about which products fell under the 1929 agreement. At Du Pont, there were advocates of the widest possible interpretation of the range of products included in that agreement as well as those who preferred to take the narrow view attributed to I.C.I. The expiration of the agreement in 1939 gave both sides an opportunity to clarify their positions. In preparation for these discussions in 1938, Du Pont's Foreign Relations Department consulted with the company's general managers who were most affected by the exchanges of information. The poll was inconclusive about the desirability of renewing the agreement. The decision to renew was ultimately made by Lammot du Pont and the Executive Committee, based on their confidence in I.C.I.'s growing research capabilities and its prospective advances in such fields as polyethylene plastics and fibers. John Jenney characterized Du Pont's attitude in 1939 as "long-time strategic rather than current tactical," in contrast to the 1929 situation when the chief concern of both the

Americans and the British had been the desire to catch up with the Germans. In July, 1939, the Patents and Processes Agreement was renewed with provisions to include new fields such as chlorine products, antiknock compounds (tetraethyl lead), synthetic resins and plastics, pharmaceuticals, and neoprene. Nylon was licensed by Du Pont to I.C.I. under a separate agreement.[3]

Both partners occasionally questioned the tangible technological gains achieved by means of the Patents and Processes Agreement during the 1930s. But their relationship was buttressed by the successes of their joint ventures. The Canadian business, C.I.L., had proved so profitable that by the early 1930s the parent companies began to scrutinize other developing countries for similar opportunities to "get in on the ground floor" of the industrialization process. Since both firms had export interests in South America, it was natural that their principal collaborative efforts were directed there.

Among the assets that Brunner-Mond brought into I.C.I. was a profitable sales branch operation in Argentina, which I.C.I. expanded in 1928 by buying into Rivadavia Company, a manufacturer of sulfuric, nitric, and hydrochloric acids that was owned in part by Bunge and Born, a large Argentine exporting firm. By 1931 I.C.I. was making plans to move Rivadavia into the heavy chemicals field. Du Pont had begun investigations along the same lines at this time, encouraged by one of its sales agents in Buenos Aires, C. J. Bartlett and Company. In the spring of 1931 Wendell Swint of the Foreign Relations Department recommended the establishment of a heavy chemicals plant in Argentina, noting the "particularly opportune" circumstances there. Despite occasional political upheavals, the country was economically sound with a large and growing urban population and a robust agricultural export trade. With the growth of eeconomic nationalism in South America, Swint predicted that the only chemical companies with any real future in Argentina would be those that had invested in local manufacturing.[4]

I.C.I.'s concurrent move into Argentina raised the prospect of possible competition, an issue not covered in the 1929 agreement except in a clause that recommended consultation between the parties when expansion was planned in "nonexclusive" territories. Swint in 1931 urged a rapid buildup of Du Pont's direct investment in Argentina in order to "put Du Pont in the driver's seat" when negotiations with I.C.I. began. His advice was taken to heart in Wilmington. By 1933, when talks began in London over a possible merger of interests, Du Pont had established two enterprises in Argentina, one in heavy chemicals and finishes; the other company, later called "Ducilo," was involved in local manufacture of sulfuric acid and imported a variety of Du Pont products, including dyes and rubber chemicals.[5]

In their discussions about a joint venture in Argentina, Du Pont and I.C.I. drew upon the C.I.L. experience as a model. They proposed to give the new company exclusive rights to their patents and processes for Argentina, Uruguay, and Paraguay, with specific limits placed on exports and on licensing outside that region. A complication was present, however, in the Argentine situation. I.C.I.'s minority partner in Rivadavia, Bunge and Born, demanded inclusion in the new venture. I.C.I. embraced a more tolerant attitude toward such local participation, but Du Pont adamantly refused to accept any third parties into the proposed merger. After much bickering, Bunge and Born's feisty owner, Adolfo Hirsch, yielded to Du Pont pressure and bowed out of the final organization of the Argentine Duperial Company in 1934. Although in this case the question of local participation was resolved in the interests of a speedy merger, the incident revealed a basic difference between I.C.I. and Du Pont in their respective approaches to foreign investment.[6]

The subsequent development of the chemical industry in Argentina highlights the degree of interdependence that characterized the international chemical fields during the interwar years. By 1938 Argentina had become an industrial battleground for Du Pont and I.C.I., local firms such as Bunge and Born, and the "big guns" of the European chemical industry, including I. G. Farben, Solvays of Belgium, and Kuhlmann of France. Two years after the formation of Duperial, Bunge and Born fired the opening salvo by announcing its intention to move into the sulfuric and tartaric acid fields. Given the limited market for heavy chemicals in Argentina at this time, Duperial approached Bunge and Born with a proposal to stay out of the tartaric acid field if the latter would keep out of sulfuric acids.

Despite Duperial's offers of "cooperation . . . in our respective independent fields," Hirsch not only rejected this overture, but proposed to expand Bunge and Born's interest in the Argentine chemical industry. In 1936, Bunge and Born, with backing from Kuhlmann and I.G., proposed to acquire La Celulosa, a pulp and paper company that had also begun to develop capabilities to produce caustic soda and chlorine through electrochemical processes. At the same time Hirsch set up a new company, Corinca, to exploit technology acquired from Kuhlmann in heavy chemicals. The prospect of Bunge and Born's merging Corinca and La Celulosa with support from major European companies posed a serious challenge to Duperial's position in the Argentine market.

Further attempts to appease Hirsch proved futile and led to dissension between Duperial's parent companies. I.C.I. was convinced that Hirsch was dangerous enough to warrant inviting his participation in Duperial. Du Pont continued to resist this step, arguing that Duperial continued to have a technical and marketing edge. Duperial had also entered negotiations with La Celulosa, but again Du Pont balked at the Argentina company's terms.

This was where matters stood in the spring of 1938 when I.G. Farben became an active participant in the negotiations. Unlike Duperial or Bunge and Born, I.G. was more interested in the pulp and paper side of La Celulosa's business. I.G.'s sudden interest was not difficult to explain. La Celulosa had begun negotiating with the Argentina government for a contract to supply nitric acid, a field which the Germans were desperately trying to control in world markets. By the summer Duperial, I.G., and Solvays had orchestrated the formation of a new company to take over the chemical side of La Celulosa's business, called Electroclor S.A. The outbreak of war in September, 1939, resulted in a reshuffling of the voting trust that managed Electroclor, with Duperial taking over the percentage held by I.G. and Solvay for the duration of the war.[7]

Despite the problems with Bunge and Born, Du Pont and I.C.I. were generally so optimistic about the new prospects in South America that even before Duperial Argentina had been launched, plans were underway for a similar joint undertaking in Brazil. Neither company had any manufacturing investment in that country, and Du Pont's sales outlet there was much smaller than the I.C.I. operation, which concentrated on heavy chemical imports. As in Argentina, Du Pont in 1934 expanded its sales activities in Brazil in preparation for the eventual merger, specializing in products such as dyestuffs, Fabrikoid, and rayon.

Duperial Brazil was set up in 1936, but the cooperation that had worked reasonably well in Canada and Argentina did not develop in this situation. The major source of difficulty was division within Du Pont over the advisability of investing in Brazil. I.C.I. advocated a substantial development of Brazil's heavy chemical industry, and Du Pont's Foreign Relations Department was generally favorable to this view. Other observers in Du Pont were more pessimistic about the prospects for rapid economic growth in that country, and these critical voices were joined by those who believed that Du Pont should concentrate on building up manufacturing capacity in newer fields such as rayon. Duperial Brazil's relatively sluggish performance in this period confirmed some of these doubts, but Du Pont was sufficiently satisfied with the success of the Argentine venture and persuaded that joint ventures ensured good relations with I.C.I., that it extended the cooperative approach to Chile, Uruguay, and Peru by the end of the decade. While these joint ventures never amounted to a substantial part of Du Pont's general business operations, they did provide the company with a foothold in markets that would become important in the future and constituted a significant element in the system of business diplomacy of the depression era.[8]

Du Pont and I. G. Farben: Allies or Adversaries?

While Du Pont's connections with I.C.I. expanded in the 1930s, relations with I. G. Farben remained at arms length, punctuated by mutual mistrust. The two companies did maintain regular communications and even entered several small-scale joint ventures while continuing desultory discussions over patent and process exchanges. But the structure of formal agreements and network of cooperative operations that characterized the Du Pont–I.C.I. relationship simply did not exist, and after 1935 neither the Americans nor the Germans evinced much interest in establishing this kind of comprehensive arrangement. As Du Pont executives reiterated throughout this period, the I.C.I. alliance, despite occasional friction, was the linchpin of the company's foreign policy. By the end of the decade Du Pont, with breakthroughs in nylon and neoprene, was increasingly confident about its technological capabilities and correspondingly less disposed to make the concessions I.G. usually demanded for access to its patents and processes.

I.G., like I.C.I., relied primarily on formal cartel agreements in its dealings with foreign companies in the 1930s. Hundreds of such agreements were made with dozens of European firms, including I.C.I., covering dyes and intermediates, nitrogen, synthetic fuels, plastics, heavy chemicals, and cellulose-based fibers. By the end of the decade I.G. had completed a complex network of arrangements with its European counterparts covering patents, prices, and markets.

Unlike the Europeans, Du Pont and other American chemical manufacturers had to be more circumspect regarding participation in international cartels. Between 1929 and 1933, Du Pont tried to straddle the fence on this issue in the field of nitrogen. This was an area in which Du Pont felt particularly threatened, since the global overproduction of nitrogen from all sources could have disastrous consequences for the company's synthetic ammonia operation at Belle. Throughout the early 1930s Du Pont representatives shuttled to and from Frankfort trying to facilitate cooperation between members of the nitrogen cartel and American producers. The task was particularly frustrating, given German intransigence and Du Pont's relative weakness in this field. By 1935 Du Pont abandoned all its efforts on this front.[9]

Meanwhile, Du Pont pursued an alternative strategy of rapprochement with I.G. through joint ventures such as those set up with I.C.I., although the results were far less satisfactory from Du Pont's point of view. The first such endeavor involved an American subsidiary called Bayer-Semesan, set up in 1928 to manufacture seed disinfectants and fungicides, owned 50 percent by Du Pont and 50 percent by Winthrop Chemical Company, which was in turn owned equally by Sterling Products Company and American I.G. Chemicals. Under the original agreement, both sides would provide Bayer-Semesan with continuous technical information, but there was an additional clause to block transfer of this information to the other parent firm, a provision of dubious practicality. By 1936 Du Pont was complaining that technical information it provided Bayer-Semesan was being passed on to I.G., and that the Germans were withholding information from Bayer-Semesan. As one Du Pont observer noted, "the I.G. is open to the suspicion of not cooperating or observing the stipulations of the contract when it does not appear to its advantage."[10]

Nevertheless, a year after Bayer-Semesan was established, Du Pont helped set up a company in Germany to produce ventilation tubes. The Du Pont investment of $31,000 in this company, called Ventube G.m.b.H., was minimal but the Americans were eager for any arrangement at this time that might conceivably lead to stronger ties between the two firms. Ventube was not a success, however, and by 1932 I.G. was seeking Du Pont's permission to dismantle the company.[11]

Du Pont's only other direct corporate link with I.G. came through Electroclor, the company established to take over the chemical business of La Celulosa in Argentina in 1938. The Germans and the Belgians each had a small share in this venture. Just as details for collaboration were being completed, hostilities erupted in Europe and Du Pont found itself the mediator between the Germans and the British, since British war regulations forbade business dealings with enemy firms. After some wrangling, I.G. agreed to withdraw from Electroclor's voting trust, but was granted an annual quota of its sales for the duration of the war.

There was one final possibility for cooperation between Du Pont and I. G. Farben through a patent and process agreement similar to the one made with I.C.I. in 1929. The Americans, despite advances in their own research capabilities, maintained a healthy respect for I.G.'s technical leadership. At the same time, Du Pont never developed the degree of trust in the good faith of the Germans that underlay its relationship with I.C.I. Between 1929 and 1936 Du Pont negotiated dozens of license agreements with I.G. in a variety of fields, including dyestuffs, rubber chemicals, refrigerants, methyl methacrylate, vinyl resins, and high-pressure synthesis processes in areas related to synthetic fibers. But all of these agreements were drafted separately and restricted to specific fields and products. The comprehensive and enduring link established with I.C.I. through the agreements in 1920, 1926, and 1929 was missing from the I.G. relationship.[12]

Furthermore, these patent exchange agreements did not embrace two areas of particular interest to both Du Pont and I.G. in the 1930s: synthetic rubber and the field of polymerized plastics and resins. Despite its own advances in these fields, Du Pont was interested in acquiring German know-how, and Fin Sparre of the Development Department doggedly pursued this objective through 1938. At first Du Pont hoped to negotiate a general agreement covering all these areas, but I.G.'s arrangements with Standard Oil (New Jersey) in the synthetic rubber field posed an insurmountable obstacle.[13]

Both Du Pont and I.G. had developed synthetic rubber by the early 1930s. In 1934 talks began in this area between the two companies, as Du Pont was concerned about the prospects of a rapid introduction of I.G.'s process, buna, into the U.S. market via Standard Oil, while I.G. hoped to acquire Du Pont's information on the development of monovinylacetylene (M.V.A.) used in the extraction of butadiene which was essential to the production of synthetic rubber through the buna process. At the same time, Du Pont hoped to use talks on M.V.A. to open the way for an exchange of patents and processes in the plastics field, where I.G. had a strong position in the development of styrene-based polymers. Before these negotiations could bear fruit, however, pressure from the German government to expand synthetic rubber production and block the transfer of chemical technology of potential military value, forced I.G. to abandon a general exchange agreement and settle for an immediate agreement on M.V.A. on the best terms possible, thus providing Du Pont with unexpected leverage. In 1938 the two companies exchanged licenses with the understanding that each party could use the information only in the production of its own version of synthetic rubber: Du Pont could use I.G.'s butadiene processes only in the manufacture of neoprene, and I.G. could use the M.V.A. process only in the manufacture of buna rubber. Acknowledgment of Du Pont's superiority was subtly indicated by the fact that only Du Pont received royalty payments.[14]

Two months after this agreement, Du Pont's desire for access to I.G. technology in the plastics field was partially satisfied by the negotiation of a cross-licensing agreement with Rohm and Haas, a Philadelphia firm that held exclusive American rights in acrylate plastics from I.G. Du Pont had begun negotiations with Rohm and Haas in 1936 to share processes in the manufacture of methyl methacrylate plastics, and in 1939 this exchange was expanded to include the plexiglass field as well. The Rohm and Haas agreement was accompanied by a broader arrangement between Du Pont and I.G. in the polystyrene field.[15]

The success of its dealings with I.G. in rubber and plastics led the company to approach the Germans in 1938 with an offer to license its new "miracle product," nylon. The Germans were very accommodating, and the only obstacle to an agreement was the concern on the part of Du Pont that the German government might provide I.G. with an unfair advantage in marketing nylon outside Germany through subsidies or related measures. This difficulty was overcome by the inclusion of a clause that would enable Du Pont to cancel the agreement in the event that there was government intervention in exports. The final agreement in May 1939 covered nylon yarn and filaments, and was supplemented by provisions in October, 1939, extending the license to include sheets and tubes, artificial leather, and wire coating.[16] The war interrupted the transfer of licenses, however, and the Americans and British, through Du Pont's 1939 license to I.C.I., retained a monopoly in the development of nylon up to the 1950s.

By the end of the depression decade, Du Pont's position vis-à-vis the Germans had changed markedly. A combination of circumstances, including Du Pont's own technical advances and I.G.'s need for access to information to satisfy the demands of its taskmasters in Berlin, led the German company to adopt a more accommodating attitude than it had in the past. Du Pont took advantage of this situation, and as a gesture of good will, made nylon available to I.G. at the same time and on terms similar to those offered to I.C.I. Whether this process would have advanced toward even more comprehensive agreements had war not broken out is hard to say. I.G.'s links with the Nazi regime in general aggravated the traditionally overbearing and secretive behavior of its leaders, although in the short run they were prepared to make concessions to erstwhile rivals. At the same time, the basic objectives of the Du Pont–I.C.I. "Grand Alliance" had largely been achieved by 1939: they had successfully defended their home markets from the Germans, and to a large extent had become the technological equals of the I.G. In contrast to the general political situation on the eve of World War II, in the international chemical industry there was a genuine balance of power.

When the issue of international cartels became a matter of intense public controversy during World War II, these arrangements with I.C.I. and I.G. were particularly troublesome for Du Pont and eventually led to an antitrust suit to break up the "Grand Alliance" in 1944. Du Pont was accused of constructing an edifice of international agreements for the explicit purpose of choking off competition at home and abroad, inhibiting American exports, and suppressing technological innovations.

These charges were not without merit, although they were exaggerated in the heated atmosphere of wartime Washington. Du Pont did deliberately seek to control the flow of technology and divert it to its own uses wherever possible. Despite the painstaking efforts made by Du Pont's legal staff to maintain a distinction between a patent exchange agreement and a cartel, the Du Pont–I.C.I. and Du Pont–I.G. agreements did have obvious commercial implications.

At the same time, these arrangements, particularly the "Grand Alliance," must be seen in the context of the unstable economic environment of the interwar period and the specific objectives of the companies. In the 1920s, Du Pont pursued patent agreements with foreign firms as part of its general strategy of diversification through acquisition of new technology. Du Pont and I.C.I. were drawn together in part because of their mutual fears of German technical leadership and commercial aggressiveness. By the end of the following decade, Du Pont and I.C.I. had developed their research and development resources to the point where they could confront the Germans on equal terms. By this time, however, the American company was linked to its British partner and other foreign companies through an intricate system of patent exchanges and joint ventures that could only be disrupted by a radical transformation of the international chemical industry. That radical transformation was shortly to come about as a result of the economic upheaval of World War II and the concurrent antitrust offensive mounted by the U.S. government not only against Du Pont but against the structure of the international industry as a whole.

CHAPTER ELEVEN

Du Pont at High Tide

Between 1929 and 1950, the United States experienced a depression of unprecedented depth, involvement in a global war on a scale far surpassing that of World War I, and emergence as the leading world power. Throughout this era, Du Pont continued to expand. Despite the setbacks of the early depression years, the full impact of Du Pont's diversification into chemicals in the preceding decade became apparent. By 1933 over 90 percent of Du Pont's earnings, including dividends from the G.M. investment, came from sales of chemical products that the company had only begun to develop at the end of World War I. Du Pont occupied a dominant position in such fields as automotive lacquers and cellophane, and was among the leaders in the industry in other products such as rayon, dyes, and synthetic ammonia. By the end of the depression decade Du Pont had not only quadrupled its 1929 sales and earnings levels, but had also begun to open up whole new fields of endeavor in plastics, synthetic fibers, and synthetic rubber. Advances in these areas helped make Du Pont essential to America's industrial mobilization in 1941–45. After the war Du Pont was able to exploit its technical achievements in domestic civilian markets, particularly in synthetic textiles, where it became and remained the leading firm for the next thirty years.

Throughout this period the dominant figures in the company were Lammot du Pont, who followed Irenee as president in 1925 and remained until 1940, and his successor, Walter Carpenter, Jr., who was president from 1940 to 1948, and chairman of the board to 1962. The two men complemented one another to an unusual degree. Despite their family connections—Lammot was Pierre's younger brother and Walter Carpenter was the younger brother of R. R. M. Carpenter and had married a Du Pont cousin—both had served long apprenticeships in the firm. Walter Carpenter had emerged as a leading figure in the Development Department during the early stages of diversification during World War I, and joined the executive committee in 1919. Lammot du Pont had begun working in the powder operations in 1907, headed the Arlington Company after its takeover in 1915, and succeeded Irenee as head of the Executive Committee in 1919. As one observer noted: "Walter Carpenter was analytical and financial-minded, orderly and logical. Lammot was practical with his production background, and had the Du Pont family common sense. Both of them had very broad points of view." Like his brother Pierre, Lammot was cautious about new investments—"he was always willing to take a risk provided he had the financial resources to pay up if he lost and if the chances were better than 50 percent in his favor"—and had a stronger interest in basic science than either Pierre or Irenee.[1] Walter Carpenter was an intensely private man, melding his public personality with the image of the company. During World War II, when the journalist Lowell Thomas wrote an article on Du Pont for the *Saturday Evening Post*, Carpenter objected to the inclusion of statements such as "Mr. Carpenter smiled," observing that remarks of this type were not pertinent to the story Du Pont wanted to present.[2]

Among the most important new directions taken by Du Pont in this period was Lammot's decision to promote basic research in chemistry. The company under Pierre and Irenee had acquired a reputation for developing technical innovations begun by others rather than initiating new processes. Irenee, in particular, had pursued a deliberate strategy of acquiring foreign technology through licensing and patent exchanges. Although Lammot continued this practice, guiding the company through the complex negotiations with I.C.I. in 1928–29, at the same time he initiated a program of fundamental research in the Du Pont laboratories. The most dramatic result of this new approach was the development of nylon in the late 1930s as a direct result of research in polymers by Wallace Carothers. By 1950, Du Pont had created a range of new synthetic fibers based on continuing research in the field, as well as a variety of related products. The company's exceptional performance in the post–World War II era derived largely from this continuing commitment to research and development.

Du Pont in the Great Depression

The depression of 1929–33 affected Du Pont, though less severely than other American companies. Total sales fell from $214 million in 1929 to $127 million at the bottom of the slump in 1932–22, and net income diminished from $78 million to $26 million, as earnings—even those from G.M.—fell by $30 million. For the first time since the end of World War I, Du Pont instituted major layoffs, reducing its work force by 10,000 in this period, while total wages were reduced by $23 million and average wage rates fell by 15 percent. Du Pont executives shared in the cutbacks: in 1929, the top executives in the company earned an average of $97,000 in salaries and bonuses. In 1932, the average was down to $35,000, primarily due to the absence of extra compensation and bonuses in that year. Lammot du Pont cut his own salary by $15,000, and practiced an ostentatious, though characteristic, frugality, by bicycling to work and reducing family spending.[3]

Despite these indications of distress, Du Pont weathered the worst of the depression in reasonably good shape. No assets of significance had to be sacrificed, and Du Pont even expanded slightly through acquisitions such as Krebs Pigments and majority control of Remington Arms. Operating investment increased as well, primarily in research and toward stabilizing costs rather than in plant expansion. By 1935 Du Pont's sales and net income were back to 1929 levels, and the number of employees exceeded the 1929 figure by 5,000, although wage rates remained marginally below the predepression level. Return on investment, Du Pont's key indicator of corporate health, doubled between 1932 and 1934, and had recovered to the 1929 level by the following year. Although the sharp slump in 1937–38 cut into sales and earnings, the drop was hardly as precipitous as in the earlier part of the decade, and Du Pont quickly recovered in 1939, buoyed by sales in new product lines. This factor, perhaps more than any other, explains the company's relatively good performance through the worst decade in America's economic history in the twentieth century. In 1937 Du Pont noted that more than 40 percent of its sales was of products not even on the market ten years before, and most of these markets experienced major growth in the period. Even in 1932, while dollar sales in fields such as ammonia and heavy chemicals were off by 60 percent, Du Pont could discern a slight increase in tonnage sales of cellophane. Furthermore, net earnings from sales in twelve new product lines grew while prices were reduced an average of 40 percent during the decade. In the Great Depression, Du Pont's strategy of diversification and its reliance on internal sources for investment proved its worth.

Du Pont's visibility as a prominent feature on the American industrial horizon and its relative invulnerability to the woes of the Great Depression, however, made it an inviting target for public ire. In 1934–35, as the company was recovering its customary economic vigor, it became the centerpiece of a Senate investigation that revived the slumbering suspicions of Du Pont as a profiteer from the misery of others. This investigation, popularly dubbed the "merchants of death" inquiry, was the byproduct of a complex set of circumstances including traditional resentment of big business; public apprehension of the dangers of war that resurfaced as the Japanese, Germans, and Italians embarked on military adventures; and, not least significant, the involvement of various Du Pont executives in the bitter partisan political battles of the New Deal era in the United States.

This last element was the result of the uncharacteristic entry of several Du Pont family members, particularly Pierre, into the hazardous world of politics; and although the company was never directly tied to these activities, the repercussions of this foray had an indirect impact on Du Pont's relations with the government in Washington.

During the 1920s Pierre du Pont and John Raskob had been drawn into the bitter debates over Prohibition, partly because of their curious conviction that the cost of enforcing this increasingly unpopular law was responsible for the continuation of federal taxes on corporate profits after World War I. By 1928 they had concluded that direct involvement in politics was the only effective course to follow to accomplish repeal of the Eighteenth Amendment. Raskob persuaded Pierre to support Alfred E. Smith, the Democratic governor of New York, and a prominent advocate of repeal, for the presidency, and to that end to use their financial resources to resurrect the bankrupt and virtually moribund Democratic party.

Although Smith was badly beaten by Herbert Hoover in the 1928 election, Raskob and Pierre du Pont maintained their control over the party's national committee, supporting another opponent of Prohibition, Newton D. Baker, Wilson's former secretary of war and otherwise an impeccably conservative Democrat, as the 1932 election approached. The Raskob-Baker faction was outmaneuvered by Franklin D. Roosevelt, however, and after the 1932 election, they lost control of the Democratic party organization. By 1934 Smith, Raskob, and Pierre du Pont were in open conflict with the Roosevelt administration, forming the Liberty League, a bipartisan organization of ultraconservative businessmen that quixotically challenged the New Deal, but was rapidly submerged after the election that year failed to erode seriously Roosevelt's control of Congress.[4]

Throughout these strange encounters, Du Pont company officials remained on the sidelines. Lammot du Pont wisely kept aloof, and other executives and family members maintained traditional contributions to the Republicans. Coleman du Pont had resigned from the Senate in 1928 and died in 1930. The Du Pont company's apolitical stance did not, however, protect it from the currents of anti–big business sentiment that surfaced at the time. Although the Roosevelt administration, which was cultivating good relations with business leaders in 1934–35, took no measures, rhetorical or otherwise, against Du Pont, western Progressive Republicans and New Deal Democrats in Congress were less restrained.

Their opportunity arose in 1934 when public interest was diverted from economic distress to the issue of international instability and the threat of war, allegedly promoted by arms manufacturers, the "merchants of death." Wars in East Asia and South America, stalemate at a disarmament conference at Geneva, and increasing tension in Europe over possible German military revival, contributed to the atmosphere of insecurity that was exploited by numerous articles and books on the arms trade. In April, 1934, a special Senate committee was established to investigate this subject, under the chairmanship of Gerald Nye, a North Dakota Republican and disciple of Robert LaFollette, the Wisconsin Progressive leader and scourge of war profiteers. The committee included stalwart bipartisan isolationists such as Republican Arthur Vandenberg and Democrat Bennet Champ Clark. The committee investigators included Stephen Raushenbush, a lawyer of Progressive leanings and Alger Hiss, who had worked briefly in the New Deal agricultural program and would become the focal point of his own controversies in future years.

Du Pont was a logical target for the committee as the virtual embodiment of the munitions manufacturer and profiteer during World War I. A parade of company officials, including Pierre and his two brothers, was haled before it during the ensuing months, and an array of charges was reviewed. For the most part, however, the investigators had to be content with rehashing the older controversies, such as Du Pont's contracts with the Allies in 1914–16, the Old Hickory affair, and Du Pont lobbying for the dye tariff in 1920–22. Du Pont spokesmen reiterated their previous counterarguments, and also pointed out that the company had largely diversified out of the explosives field: the Smokeless Powder Department had received the smallest share of new investment of all Du Pont's operations and had contributed only about 2 percent of the company's total earnings in the years 1923–33.

The investigators made more headway with Du Pont's connections with Remington Arms that had sold weapons to both sides in the Chaco War in South America in 1931–33, and its links with foreign arms makers, including I.C.I. and the German explosives companies. Du Pont responded that the Remington takeover was intended principally to give the company a foothold in the sporting ammunition market. Companies like Remington that produced hunting rifles were threatening to integrate back into the powder field, hitherto a Du Pont preserve. Through its control of Remington, Du Pont could hold them at bay. But the Wilmington firm had more problems explaining the foreign business connections. This was the first general public airing of the intricate arrangements with I.C.I. Du Pont executives argued that military explosives constituted only a minor part of these agreements and that Du Pont had stopped exchanging patents and processes in the explosives field in 1928 (although it had provided Mitsui of Japan with information on the synthetic ammonia process in 1931); but the impression lingered that the company maintained dubious contacts with foreigners.

On the whole, however, Du Pont emerged from the ordeal with little damage to its reputation and no threats to its operations. Irenee du Pont was characteristically disdainful and outspoken, ostentatiously puffing smoke rings for photographers during the hearings, and suggesting the need for an "absolute monarch" during wartime. Pierre, Lammot, and lesser officials were more restrained and clearly did not fit the stereotype of demonic arms profiteers. In any case, the committee investigation began to lose momentum in 1935. Proposals for nationalization of the munitions industry never got off the ground, and although the Nye committee hearings contributed to the passage of the Neutrality Act of 1935 that was rapidly abandoned after World War II began, no other substantial legislation proceeded from its reports.[5]

From Du Pont's point of view, however, the hearings did have some pernicious results. Attention had been drawn to the I.C.I. agreements and, even though no one contested their legality at the time, antitrust lawyers in Washington would find them a fruitful field for probing, eventually mounting a full-scale and successful assault in 1944–51. The munitions inquiry had also resurrected Du Pont's old image of war profiteer that the company had hoped to replace with a more benign appearance. Lammot du Pont purportedly considered pulling the company out of military explosives altogether but was dissuaded from doing so by U.S. military officials who foresaw the need for Du Pont's immense reserve capacity in wartime.

One direct result of the Nye hearings was a decision to promote public relations. Other companies, notably Standard Oil (New Jersey) and A.T. & T., had pioneered in this area in the previous decades, but Du Pont, as a company largely serving industrial buyers, had been relatively indifferent to its public image. In 1936, however, the company embarked on a program of "institutional publicity," including films on the chemical industry; exhibits at the various commercial fairs of the late 1930s that climaxed with the New York World's Fair and San Francisco Golden Gate Exposition in 1938–39; and sponsorship of a radio program, "The Cavalcade of America," that broadcast for almost twenty years and was endorsed by the American Legion as "the program most acceptable and worthwhile to the family audience."[6]

In one area of corporate sensitivity, Du Pont had remarkably few problems in this era. The depression decade was punctuated by increasing tension between business and organized labor as the C.I.O., buffered by the Wagner Act and related government measures, sought to organize the mass production industries that had successfully resisted unionization in the past. The steel and auto industries, in particular, witnessed bitter confrontations: General Motors was involved in one of the most publicized of these episodes, the sit-down strike at the Flint body works in 1937. Du Pont, however, avoided this conflict and indeed remained largely nonunionized up to the present. In 1981, before the merger with Conoco, fewer than 5 percent of Du Pont's 60,000 wage workers were members of international unions despite persistent efforts at organization by the United Steel Workers for eight years.

Like other large chemical firms abroad, Du Pont had traditionally followed paternalistic practices toward its employees, reflecting the persistence of family control of the firm. In the nineteenth century, workers in the Brandywine mills lived in Du Pont–owned houses, and their children attended Du Pont-sponsored schools and churches built by descendants of the agnostic Pierre S. du Pont de Nemours. Victims of the frequent accidents in the powder mills or their dependents received compensation and pensions. This kind of proprietary regime proved harder to maintain, however, once the company began to expand, and the managers of the reorganized corporation after 1903 attended to labor relations in their customary systematic fashion. Characteristically, the innovations in this area constituted a form of updated paternalism that provided benefits to workers without diluting management control over the workplace, an approach labor historians have labeled "welfare capitalism." In 1909 a regular pension and compensation plan was established for workers with more than fifteen years service with Du Pont, to supplement the public workmen's compensation. At the same time, the company introduced a Stock Investment Plan under which employees were encouraged to purchase Du Pont debentures. In 1919 Du Pont set up group life insurance for its employees, extended in 1929 to include sickness and accident coverage. For its senior executives, Du Pont paid out stock bonuses and in 1927 established a trust fund that provided executives with low-interest loans to exercise stock options. None of these measures was unique to Du Pont, but the company was very shrewd in distributing its benefits to promote company loyalty.

In 1921, after decentralization was introduced, Du Pont initiated an Employees Representation Plan, establishing joint labor-management councils at sixty-seven of its plants. Here again Du Pont was following a practice adopted by other large corporations, notably Standard Oil (New Jersey) in this era, but the Du Pont "company unions" proved to be exceptionally durable. In 1934 the company ended management participation in the councils to enable them to qualify under section 7(a) of the National Industrial Recovery Act. Later in the decade, when company unions in other firms, including G.M., were displaced by independent unions, Du Pont easily rebuffed its only serious challenge, a United Mine Workers attempt to organize the rayon plant in Belle, West Virginia.

Du Pont's success in this area can be attributed in part to its general growth during the depression period and relatively high wage rates. By 1936 Du Pont's total employment exceeded 1929 levels and wage rates were 12 percent higher than in the predepression period. Total wages and salaries consistently exceeded net income from operations, since Du Pont had the cushion of earnings from G.M. and other investments and rarely had to face a choice between shareholders' interests and those of employees. Furthermore, the company continued to expand its benefit program, providing wages for up to three months' for nonoccupational disabilities, extending long-term disability payments to workers with only one year's service, and establishing a research organization to investigate potential health hazards in the plants, and to improve safety conditions, a belated response to the Deepwater incident.

But beyond these concrete benefits, Du Pont was able to inculcate a sense of corporate loyalty that persisted over the years. This sense of identification was probably stronger among skilled workers and chemical specialists than others, and more evident at Du Pont in the period before World War II than in the postwar era when the company became increasingly committed to large-scale synthetic fiber production. This factor is, however, important in Du Pont's labor relations. Du Pont employees did not just have a job; they had a career. The company's elaborate system of organization established a clear connection between performance and advancement, and a rational structure of rules and procedures. Discipline was neither whimsical nor overtly oppressive. In practice this system may have had shortcomings, but it worked sufficiently well to encourage productivity and alleviate the dissatisfactions that have created unrest in other large industrial organizations in the 1930s and in the more recent past.[7]

The Nylon Revolution

Through these years, while Du Pont experienced the upheaval of depression and the attacks of senatorial inquisitors, the groundwork was being laid for a transformation of the company as dramatic as that which occurred in 1916–22. Between 1929 and 1939 a procession of new products and processes emerged from the Du Pont laboratories, including synthetic camphor, synthetic urea used in plastic production, hydrocarbon resins used in printing, and "Lucite," a transparent plastic later adopted by the aircraft industry. This continuing diversification of products contributed to Du Pont's buoyancy in the depression years, but, more important, set the stage for a renewed surge of development during and after World War II.

Two new developments in particular that would contribute to Du Pont's growth appeared in this era: nylon and neoprene. Both were associated with the work of one man, Wallace Carothers, and were the direct result of Lammot du Pont's commitment to basic research. Du Pont's interest in applied research in industrial chemicals was well established. Since 1903, with the foundation of the Wilmington experimental station, the company had promoted research, concentrating it within the Chemical Department in 1911, which in turn worked closely with the Development Department in the initial diversification projects. In 1921 research was decentralized to encourage each division to maintain its technical edge, but the Chemical Department's director retained overall coordination of research operations.

Irenee's strategy of diversification through acquisition of foreign patents had proved successful in the 1920s, but Du Pont scientists were anxious to move beyond derivative activities. When Lammot assumed the presidency in 1925 he was approached by Dr. Charles M. A. Stine, the chemical director and one of the early veterans of the Repauno labs, who urged him to support fundamental research, recalling the tradition of Lammot's father and namesake who had developed soda powder and founded Repauno. Lammot agreed and set up a new wing at the experimental station, promptly dubbed "Purity Hall" by its neighbors.

Stine detailed Dr. Julian W. Hill to help recruit researchers for Purity Hall. Carothers, the most important of these recruits, was lured from Harvard University in 1929, attracted less by the salary than by the offer of virtually complete freedom and substantial material assistance in carrying out his research.

Carothers proposed to pursue research along a path recently opened by the German chemist Hermann Staudinger in the study of large molecular chains, called polymers. Throughout the preceding century chemists had been largely preoccupied with the reduction of natural materials into their constituent elements and the analysis of these simple molecular structures. The analysis of more complex substances, such as cellulose, starch, and rubber, posed greater difficulties since these materials when broken down could not be recombined. Although materials based on complex molecular structures could be used in industrial chemistry—cellulose was a major raw material for a wide range of products, for example—the process through which these structures were formed remained unknown.

In 1907 Leo Baekeland, a Belgian chemist who emigrated to the United States, developed the first commercial synthetic polymer, combining phenol and formaldehyde into a viscous material that could be used as a lacquer. Earlier chemists, including Adolf von Baeyer of B.A.S.F., had produced this material in their labs, but its potential uses had been ignored. Baekeland marketed his material as Bakelite, one of the first commercial plastics; but the real significance of his work was that he had created an entirely new material, a chain of molecules from simple compounds in a structure that had never existed before.

Baekeland's achievement stimulated greater interest in the structure of complex substances, whose analysis was augmented by advances in concepts of atomic structure and technical equipment such as X rays for investigating the behavior of atoms. These advances undergirded Staudinger's argument in 1926 that complex organic substances are composed of long chains of simple molecules combined in specific repetitive patterns which could be replicated in the laboratory.

The implications of Staudinger's theory were not lost on the chemical industry. If giant molecules could be built from simple ones, as Baekeland had formed his plastic, then virtually any organic material could be synthesized from simple and readily available sources. The basic challenge was to discover the processes through which the giant molecules could be constructed.

Carothers had already begun research on the subject of molecular bonding when Du Pont approached him. Even though the company had no idea in which direction or how successful Carothers's research would be, there could be little doubt of the commercial potential of the polymer field. Furthermore, Du Pont was not alone in moving into this area. In 1927 I. G. Farben set up a polymer research laboratory that had already by 1929 synthesized polystyrene, and was moving rapidly forward into the field of plastics, which it would dominate in the next decade as it had dominated dyestuffs before World War I.

Carothers's main focus was on the process of polymer construction, but the commercial byproducts of Purity Hall's research appeared fairly rapidly and regularly. During the decade after its establishment, the polymer research team contributed over fifty patented materials to Du Pont, beginning with neoprene in 1931. This was one of the earliest synthetic rubber substances, developed from polymerized acetylene treated with acid. Neoprene soon found a market as a sealant, since it was more resistant to water, acids, and oils than natural rubber, but the process was initially too expensive to provide a substitute in the largest field of rubber application, automotive tires.

The search for a synthetic rubber suitable for this latter use was being pursued with intensity not only by I. G. Farben but also by Standard Oil (New

York) as an offshoot of its work in petroleum cracking and refining. The development of a synthetic rubber had an immediate commercial impetus in the 1920s when the British rubber cartel was restricting supplies, but its complicated and irregular molecular structure, which contributed to rubber's elasticity, made duplication difficult. In 1928 the Germans were able to construct a rubberlike polymer from isobutylene, but it lacked the quality of the natural material. Not until the mid-1930s, when Standard Oil developed a process for extracting butadiene from oil, was the technical capability available for the manufacture of synthetic rubber for tires.[8]

Meanwhile, Carothers turned his attention to the development of a group of polymers called polyamides, plastic materials that could be molded when heated but would then assume a permanent shape. In 1930 Hill had produced, more or less accidentally, a plastic that exhibited many of the qualities of cellulose fiber: tough, elastic, and somewhat heat-resistant. Over the following years Du Pont researchers experimented with a large variety of polymerized fibers before finally developing one that combined the properties required for commercial-grade fiber, from hexamethylenediamine and adipic acid. This material could be extruded like rayon and other cellulose-based fibers but was produced entirely in the laboratory from simple chemical ingredients, could be drawn more tightly into yarn than rayon, and was much stronger than silk or cotton.

In 1937 Carothers took out a patent on the polyamide and Du Pont began to move toward commercial production. After considerable deliberation the fiber was given the otherwise meaningless name "nylon," presumably to indicate its market relationship to cotton and rayon. Nylon had many potential uses beyond the textile field. In 1938 it was initially marketed in toothbrushes, electrical wire insulators, violin strings, and assorted unrelated products. A pilot plant was established at Seaford, Delaware, at an initial investment cost of $8.6 million, and the Belle rayon plant was expanded to supplement Seaford's four million pound capacity.

Carothers did not live to see the commercial development of his research, which aspect had never particularly interested him. Depressed by the death of his sister and by an unhappy marriage, the chemist committed suicide shortly after filing the nylon patent. His research team, however, continued to pursue his work on polymers. Du Pont had spent, by its own estimate, $6 million on the research effort in the preceding decade and it was now poised to exploit the results.[9]

Probably the most publicized use of nylon in its early years was in women's hosiery. The business press gleefully reported the huge lineups for synthetic stockings when they were first marketed in May, 1940. Within two years the production of nylon hose had quadrupled, but at this point America's entry into World War II virtually ended civilian nylon production, since the military purchased the material for parachutes, surgical sutures, and a wide range of other uses, including airplane tire cord.

After the war Du Pont expanded its nylon capacity with four plants in operation by 1948, by which time the company had estimated annual sales in this field of $38 million against an $84 million investment. Meanwhile, Du Pont had embarked on the development of other synthetic fibers: orlon, a material with woollike properties that was quick-drying, went into production in 1947; and "dacron" polyester, a polyethylene-based material that had been acquired by I.C.I. from its British creators during World War II and licensed to Du Pont, which had not included polyesters in its nylon patent.

New processes also proceeded from continuing research in polymers. One of the most significant of these plastics was developed by accident by Dr. Roy Plunkett in 1943 while he was doing research on fluorocarbon refrigerants. A material composed of tetrafluoroethylene polymerized in cold storage and subsequently proved to be exceptionally heat resistant, anticorrosive, and impregnable to mold. The plastic, shortened to "teflon," was used as a coating in the casing of the first atomic bomb. Subsequently, Du Pont chemists copolymerized teflon with other large molecules to produce coatings with a wide variety of uses in tapes, raincoats, automobile and airplane exhaust systems, and nonstick pots and pans.

Nylon remained the single largest contributor to Du Pont's earnings from operations during the two decades after World War II. Under Crawford Greenewalt, Carpenter's successor as president, Du Pont focused its efforts on exploiting the nylon market. In certain respects this may have been a mixed blessing over the longer term for Du Pont. In the early 1930s Du Pont had been a truly diversified company: no one product area accounted for more than 15 percent of sales, and the spread of products among industries as diverse as automobiles, construction materials and textiles was relatively even. Increasingly after World War II Du Pont emphasized synthetic fibers: in 1947 these accounted for one quarter of all sales and by 1957 more than one third. No other product line approached it in volume. To a large extent the Du Pont of the 1950s had become the "nylon company," and while the earnings remained substantial, the company was more vulnerable to shifts in the fiber markets and the growth of competition, particularly after Du Pont began licensing nylon to Chemstrand Corporation, a subsidiary of Monsanto and American Viscose, in 1950. To be sure, Du Pont continued to promote new research—to the point where Greenewalt was criticized in some quarters for overemphasizing research in areas whose commercial benefits were not readily apparent—and developed more diversified application for its synthetic fibers and plastics. The results of this growing concentration in the fiber field would consequently not become evident until well into the 1960s.

All these developments lay well in the future as Du Pont and the American economy emerged from the depression. Over the next twenty years the transformation of Du Pont and the chemical industry symbolized by the development of nylon would carry it through its most prosperous era with net sales increasing almost six times over, net income tripling, and overall return on operating investment rising from 9 percent to 13 percent.

From World War II to Korea

By the time war broke out in Europe in the autumn of 1939 the bitter political disputes of the middle of the decade had apparently faded into the background. The Du Ponts and the Roosevelt administration had generally toned down their mutual rheotorical attacks, and public concern over the "merchants of death" had been largely displaced by more deep-seated fears of the bellicose nations of Europe and Asia. In 1940 Du Pont began constructing smokeless powder plants for the United States and British governments on a fixed-fee basis, and agreed to provide royalty-free licenses on a number of its patents to the government.

Over the next four years Du Pont built twenty-one smokeless powder plants and assisted in the construction of another nineteen, producing over two million pounds of powder for the U.S. military forces, about 55 percent of total U.S. production. In addition, the company developed improvements to neoprene, contributing 7 percent to the country's crash program in synthetic rubber production, and rayon yarn for tires as well as nylon for parachutes and other uses. Total dollar sales rose from $400 million in 1940 to $962 million in 1944, with sales to the government accounting for 22 percent of the total. Du Pont carried on its payroll between 40,000 and 50,000 employees in government plants in addition to between 62,000 and 67,000 in its own operations in 1944–45.

The most significant of Du Pont's wartime contracts with the government was negotiated in December, 1942, when the company agreed to construct a pilot plant at Oak Ridge, Tennessee, followed by a larger works at Hanford, Washington, to produce plutonium for the Manhattan Project. The initial contract between Du Pont and the scientists working on the atomic bomb came through Charles Cooper, a chemical engineer detailed to work with Arthur H.

Compton of the Metallurgical Laboratory at the University of Chicago earlier that year, on the development of a chain reaction in the uranium pile accumulated by the Met Lab. Compton recommended to James Conant of the Office of Scientific Research and Development, who had a long-standing acquaintance with Charles Stine of Du Pont, that the government retain Du Pont to construct a plutonium separation plant.

Compton's suggestion was enthusiastically endorsed by General Leslie R. Groves, the Army engineer who took charge of the Manhattan Project in the summer of 1942. Groves had worked closely with Du Pont in constructing powder plants, and he stressed the fact that Du Pont had its own engineering department to undertake plant construction as well as long experience at managing complex research organizations. In November, after Du Pont was approached, Stine and Bolton went to Chicago to observe Compton's operations.

Du Pont's initial reaction to the government's proposal was characteristically guarded. The company had no experience in the field of atomic research, and a commitment to the Manhattan Project would draw heavily on a research staff already committed substantially to other wartime projects. Stine and Bolton were uncertain whether Compton's experiments would be successful. Nevertheless, the company agreed the following month to undertake construction of a chemical separation plant for plutonium recovery. Under the contract the government would underwrite all costs of construction and operation, which ultimately came to more than $390 million. Du Pont, still mindful of its experience with the "merchants of death" inquiry, asked only for a fee of $1. In addition, the company arranged for a $10.5 million insurance policy for the venture, noting the novel and possibly extreme hazards associated with the operation.[10]

Relations between Du Pont and the scientists involved in the Manhattan Project were not always smooth. Habitually cautious about new developments, Du Pont resisted Compton's desire to move directly from experimental operations to full-scale plutonium works and insisted on establishing a pilot plant at Oak Ridge that was under construction in February, 1943, and completed five months later. Du Pont also prevailed on Compton and the Met Lab to assume full responsibilities for operating the Oak Ridge plant, which the scientists were reluctant to undertake. Later observers also commented on the tensions that sometimes surfaced between Du Pont engineers, accustomed to a disciplined and structured work environment, and the more free-wheeling researchers from the universities.

Despite these differences, the project as a whole moved forward steadily through the next two years. Construction of the large plutonium extraction works at Hanford began in the fall of 1943 and was essentially completed within a year. At its high point Hanford employed over 50,000 people, including 6,000 managers and scientists. The operation was carried out under E. B. Yancey of Du Pont's Explosives Department, with a staff drawn from many of the other departments. Among the most important of Yancey's aides was Crawford Greenewalt, who had worked for a time with Carothers on nylon development and rose to research director at Du Pont's heavy chemicals (Grasselli) Department at the time Du Pont became involved in the Manhattan Project. Greenewalt had been a participant in the discussions leading up to Du Pont's commitment in the fall of 1942 and in December of that year was a witness to the first nuclear chain reaction carried out by Compton's Met Lab in Chicago. Greenewalt's work at Hanford made him a highly visible and respected executive in the company—abetted by his marriage into the Du Pont family in the 1930s—and helped lay the groundwork for his rapid rise to the presidency of Du Pont in 1948 when Carpenter moved up to become chairman of the board.

When the war ended Du Pont was anxious to extricate itself from Hanford. Some consideration was given the prospect of a continued commitment, since Du Pont engineers believed the efficiency of plutonium production could quickly be upgraded. But as Stine pointed out, this move would involve a major entry by Du Pont into the electric power field "quite removed from the company's present line of business." Du Pont was persuaded to continue to operate Hanford until November, 1946, but at that point terminated its contract with the Atomic Energy Commission. Two years later, however, the A.E.C. requested Du Pont to undertake further work on plutonium recovery, and in 1950 the company entered a new contract to construct a nuclear fuel plant at Savannah, Georgia, part of the government's crash program to develop a thermonuclear bomb.[11]

Government contracts benefitted Du Pont, as they benefitted many of the major American industrial corporations during World War II and the remobilization of the economy in the Korean conflict. The war economy did not have as dramatic an effect on Du Pont in this era, however, as it had in 1914–18. Du Pont's net income after taxes during 1938–46, including earnings from its G.M. investment, came to $725 million, of which the company retained $118.7 million for reinvestment. New investment costs in this period came to $486 million, so that retained income contributed only one quarter of the requisite capital, in contrast to the World War I era when earnings from operations contributed more than 75 percent to Du Pont's expansion. Furthermore, during the war years 1941–45 Du Pont's average net income, exclusive of G.M. earnings, was $76 million per year, while in the postwar period 1946–50 before the Korean War, average net income leaped to $182 million per year. Sales to the civilian economy, particularly of synthetic fibers that contributed almost 30 percent of the total by 1950, provided the most profitable sources of revenue. Sales grew at an average annual rate of 13 percent in the decade following the end of World War II, and average return of investment ran at 11.4 percent per year except during the Korean War period of 1951–53.

While government largesse appeared less significant to the company's growth and earnings in the wartime and postwar years, Du Pont executives regarded some government policies as positively inimical to the company's well-being. Particularly obnoxious were the excess profits taxes imposed during World War II and reactivated in the Korean War. In 1941 Congress raised the rates on both "normal" profit rates (defined as 95 percent of average net income for the years 1936–38 plus 8 percent net capital addition) from 24 to 31 percent and on "excess" profits from 50 to 60 percent. In the following two years the excess profit tax rate was jacked up again to 90 and finally 95 percent. Subsequently the law was amended to allow companies subject to excess profits taxes to apply for a postwar credit of up to 10 percent of taxes paid under this provision; but the government also required a renegotiation of war contracts after their cancellation to recover estimated value of capital expansion. Du Pont set up a special reserve fund to cover these anticipated repayments that came to $3.3 million. During the war years the tax rate on gross revenues doubled from prewar levels, averaging 14 percent between 1942 and 1945. Du Pont preferred to consider the tax burden in the context of net income, noting that taxes amounted to almost twice net income in 1943–44, and one and a half times net income in 1945, although the company was later able to recapture part of the total in tax credits.

Du Pont's fulminations against taxes were relatively restrained during World War II, compared with Pierre's outbursts following the passage of the first excess profits tax law in 1916. Nevertheless, the company joined in the chorus of complaints about excess profits taxes that led to the rapid dismantling of wartime tax measures in 1945.

When the government reimposed an excess profits tax during the Korean War, business opposition surfaced more rapidly and remained more vocal than in the earlier wartime situation, when government appeals to patriotism had more force. The excess profits tax of 1950 was in certain aspects more generous toward business than preceding laws, allowing a substantial tax credit based on earnings in the 1946–49 period (when earnings were of course significantly higher than in the previous base period of 1936–38), and imposing a maximum tax rate of 62 percent on excess profits through a combination of surtaxes. But efforts by the Truman administration to increase rates in 1951–52 were blocked by Congress, which was restive over the Korean stalemate and buffetted by antitax lobbies from labor as well as business organizations. Du Pont spokesmen asserted that the excess profits tax "imposes a specific penalty on growth and efficiency by taking away most of the earnings from new operations and from technological improvements," with rates as high as 82 percent on earnings from new plants and processes. The removal of the Korean War taxes, along with wage and price controls, was one of the first orders of business for the incoming Eisenhower administration in 1953.[12]

Rising taxes bothered Du Pont leaders in this era, but these burdens were partially offset by a general rise in the volume of business generated by government spending, and the tax rate only became a serious problem for a short time during the war emergencies. More troubling for Du Pont was the steady and implacable advance of antitrust authorities, who seemed impervious to political changes in Washington and intent on probing and ultimately prosecuting Du Pont for virtually every aspect of its business operations. The antitrust offensive peaked in the years 1948–57 with major litigation involving Du Pont's links with I.C.I. and National Lead, its control of the cellophane market, and, most distressing, its investment in General Motors. The struggle to retain its existing structure would preoccupy Du Pont for much of the next decade and, despite some successes, the struggle proved to be in vain and Du Pont was perforce transformed once again. Paradoxically, however, if one accepted the lamentations of Du Pont officials and attorneys at the time, the company emerged from this ordeal with little permanent damage and in certain respects a stronger, more coherent organization. Du Pont rapidly developed new strategies not only to compensate for the changes forced upon it but also to take advantage of new opportunities that were emerging, for the international chemical industry was undergoing a transformation no less dramatic in this era as political, technological, and commercial factors converged to create a truly multinational business community.

CHAPTER TWELVE
The End of the Cartels

For decades prior to World War II the international chemical industry had successfully practiced a refined form of business diplomacy, relying on a strong code of industrial cooperation to ensure order in world chemical markets. During the interwar years this cooperative ethic had reached a peak, with regulated competition becoming the rule rather than the exception in the chemical industry. Ever conscious of the U.S. government's erratic pursuit of antitrust violators, Du Pont had straddled the rather uncertain line between agreements with European chemical companies and the foggy boundaries established to protect American competition under the Sherman Act. By the time war broke out in September, 1939, however, Du Pont's intricate foreign policy had begun to unravel.

As early as 1938 the company perceived a marked change in the Roosevelt administration's attitude toward business combinations, underscored by the appointment of Thurman Arnold to head the Antitrust Division of the Justice Department. When it became apparent that the government might take a closer look at Du Pont's foreign arrangements, the company dispatched a mission to London to try to obtain modifications in the I.C.I. agreement that would circumvent any novel interpretations of the antitrust laws. Perhaps just as important as the external pressures being brought to bear on Du Pont in the late 1930s were the internal sources of friction between Du Pont and its global counterparts, particularly I.C.I. By the time the United States entered the war this combination of internal and external pressures had seriously eroded the "Grand Alliance" developed over the preceding decades.

The "Grand Alliance" in Decline

By the 1930s the clearly restrictive aspects of the Du Pont–Nobel alliance such as provisions for divisions of markets, sales quotas, or production limitation had been supplanted by agreements which concentrated on technological exchanges and joint ventures in foreign markets. Undeniably, the more cartellike ties between the partners continued to be critically important for the British chemical company, whose overseas sales and production dwarfed Du Pont's international operations. As long as Du Pont could be certain that its home market would not be inundated with European chemical products, the international cartelization of the chemical industry did not dominate their strategic thinking. Indeed, by the mid-1930s the unstable markets and chaotic prices within the industry that had historically prompted Du Pont to seek ties with the European companies had become much less important among the factors entering into its plans for expansion. Instead, a new cluster of developments made it appear desirable for the company to ease out of its overseas agreements and entertain the possibility of a more independent course abroad.

Despite the generally successful performance of the partnership, particularly through the joint ventures such as C.I.L., there were some underlying problems that became more pronounced in the late 1930s. For one thing, I.C.I.'s competitive position outside its protected markets was deteriorating. In the Far East, particularly in China, I.C.I. had to face rising Japanese exports, while in South America the Germans and even Italians challenged Britain's traditional position. Even within the British Empire, pressures for local manufacturing posed a threat to I.C.I. exports. In the latter situations, of course, the British company had developed a practice of preempting local competition by undertaking direct investment, as in Canada, South Africa, and Australia. But this strategy had its drawbacks. Direct investment overseas in the depression era reduced the volume of manufacturing at home, which could prove politically hazardous even though the returns from most foreign ventures were considerably higher than from goods produced within the British Isles. Furthermore, as G. P. Pollitt noted in 1935, "in each country in which we trade, we are finding ourselves less and less able to sell goods manufactured in Britain and more and more compelled to look upon our . . . foreign companies . . . as manufacturing concerns from which we derive an investment revenue only"—one that could eventually be blocked from repatriation by the imposition of local exchange controls and similar measures, in which case these earnings "must inevitably be finally lost, as there will exist no way of bringing it back."[1]

Foreign direct investment also could lead to further complications, as the local manufacturers began to press I.C.I. to allow them to export their own surplus production. This problem was already apparent in Canada, where C.I.L.'s aggressive chief executive Arthur Purvis several times endeavored to export products not explicitly restricted under the 1929 Du Pont–I.C.I. agreement, asserting that the Canadian government was pressing him to increase sales abroad. Purvis's forays beyond the Canadian border became the source of considerable acrimony with Du Pont and led to a general restructuring of the tripartite agreement in 1936.[2]

I.C.I.'s problems in its own markets did not cause a deterioration in the relationship with Du Pont—the "alliance" had always been based on mutual considerations of defense against aggressive competitors—but did reflect the changing circumstances of the two firms. Du Pont in the 1930s was transforming itself into a formidable technological dynamo, no longer dependent on foreign patents and processes to guarantee its security. The American company still respected the traditions of gentlemanly cooperation with its international counterparts, but to an increasing degree Du Pont executives such as Walter Carpenter emphasized the context of "enlightened self-interest" in which such agreements were observed. Furthermore, Du Pont began to adopt a more suspicious attitude toward I.C.I.'s relations with other firms, based on its recognition of the interrelations of new developments in chemical technology, while at the same time taking a tougher bargaining stance with I.C.I. during discussions of new patent and process exchanges in the late 1930s.

The subtly shifting relations of the two companies were revealed in the course of negotiations over two significant technical developments, Du Pont's nylon and I.C.I.'s polythene, between 1937 and 1942. In January, 1937, Lammot du Pont wrote to H. J. Mitchell, president of I.C.I. (McGowan was out of town on a tour of the company's domains) informing him of Du Pont's development of a noncellulose fiber that was pointedly described as a "major invention" which under the 1929 agreement would require special arrangements, meaning that Du Pont would not automatically release the technical information to I.C.I. Mitchell countered with the claim that I.C.I. too had a "promising product," an artificial wool called "Ardil," derived from vegetable proteins.

A little more than a year later, serious negotiations began, by which time Du Pont had already made nylon licensing agreements with a French company, Rhodiaceta, and an Italian firm, Societa Elettrochimica del Toce. To the surprise of I.C.I., Du Pont introduced a third party to the negotiations in the person of Sir John Hanbury-Williams, a director of Courtaulds, the British rayon manufacturer. I.C.I. was not averse to working with Courtaulds, since the British chemical company, unlike Du Pont, had no previous experience in the textile field. Nevertheless, Du Pont's initiative was unusual, particularly since only a year earlier I.C.I. had encountered "rumblings" from Wilmington when it negotiated the sale of a methyl methacrylate resin patent to Rohm and Haas in Germany, whose subsidiary in the United States had an exchange agreement with Du Pont.

After several months of discussions, a nylon exchange agreement was completed in July, 1939, according to which Courtaulds and I.C.I. would undertake a joint venture with Du Pont's licenses. But I.C.I. was less than completely satisfied with the outcome. The British company received an exclusive license for the Empire, but the conventional provisions of the 1929 agreement that would have given I.C.I. a nonexclusive license in other territories, enabling it to enter the European market, was withheld, since Du Pont proposed to make similar exclusive licensing arrangements with I.G. and the French and Italians. Only after much pressure, including intimations that the British government supported I.C.I.'s positions for reasons of national security, was Du Pont persuaded to extend to I.C.I. a nonexclusive license in the Baltic countries and Portugal.[3]

While Du Pont and I.C.I. were engaged in nylon discussions, the British company had been developing polythene, a polymer-based plastic that had exceptional insulating qualities and helped provide British radar with the substantial advantages that contributed to success in the air battle with Germany during World War II. I.C.I. regarded polythene as important a breakthrough as nylon and hence a "major invention." After some equivocation, Du Pont accepted this view when approached by I.C.I. in March, 1939. Du Pont sent a mission to observe the operations at I.C.I.'s Winnington plant where polythene was being commercially developed, but did not pursue the matter much further until May, 1940, when Crawford Greenewalt informed I.C.I. that Du Pont "had been carrying forward experimental work . . . in the field of high pressure polymerization." About eighteen months later, after McGowan formally offered to share the polythene process with Du Pont for wartime use, with postwar terms to be settled later, a second Du Pont group arrived in England and "took back with them a huge volume of reports and drawings and a list of information which we required from them," according to an I.C.I. observer, but which was very slow in coming. I.C.I.'s polythene specialists were convinced that Du Pont "had been working behind I.C.I.'s back with the object of bringing down the rate of royalty which they would have to pay on any license they might take out."[4]

Senior executives, including McGowan and Carpenter, intervened at this point to restore friendly relations, but the polythene dispute, like the nylon licensing controversy, was a portent of things to come when the "old guard" at both Du Pont and I.C.I. finally departed from the scene. The alliance was also put under strain as a result of the wartime situation. During the fall of

1939 I.C.I. tried to enlist Du Pont aid in the British government's attempts to squeeze German exports out of South American markets. Given the shortage of chemical products available for export, I.C.I. asked Du Pont to "adopt the magnanimous attitude of supplying these goods yourselves in the interim." I.C.I. would naturally resume its position in these markets as soon as possible after the war. Mindful of possible problems with American neutrality, not to mention antitrust laws, Du Pont declined, and its refusal to place Britain's strategic requirements above its own "sound business considerations" led to further irritation in I.C.I.'s ranks over their transatlantic partnership.[5]

Despite the tensions aroused through the years of depression and war, the "Grand Alliance" might well have survived and prospered during the postwar years had it not been for the simultaneous pressure brought to bear on the relationship by the American government. Beginning with an ambitious investigation of the concentration of economic power in American society undertaken by the Temporary National Economic Committee (T.N.E.C.) in 1938, American firms doing business abroad, exchanging patents with foreign firms, or engaging in cooperative activities with European industries were continually in the limelight. International cartels became the nation's scapegoat for a variety of problems ranging from the Great Depression to unpreparedness for the struggle against the Axis. Du Pont and I.C.I. were just two of dozens of companies examined by probing officials connected with T.N.E.C., the Antitrust Division of the Justice Department, and a series of Congressional investigations that continued through the World War II era.

Although the T.N.E.C. investigation concentrated on domestic industries, the connections between American firms and foreigners through such mechanisms as patent exchanges were also brought under consideration. This subject aroused the interest of Thurman Arnold, the newly installed assistant attorney general for antitrust, who initiated the novel practice of clustering antitrust cases in specific areas so that the cases, taken as a whole, revealed more precisely patterns of restraint of trade in American industries. Between 1939 and 1942 Arnold's staff meticulously gathered a wealth of data on industries such as chemicals where Arnold believed that abuses of the patent laws could be demonstrated. To buttress his legal position, Arnold did not hesitate to make dramatic appearances before congressional committees and distribute public pronouncements that directly linked cartels with Nazi war preparations, implying that American firms that had participated in arrangements with European firms were both greedy and unpatriotic. These techniques contributed to Arnold's reputation as a trustbuster—more antitrust suits were filed during his tenure at Justice between 1938 and 1943 than in the preceding forty-five years—and although his record in court was not quite as impressive, Arnold proved to be an effective antitrust chief by persuading many companies to settle potential cases through consent decree proceedings. Arnold's successor, Wendell Berge, a less flamboyant but more tenacious enforcer of the law, preferred to push cases through the court and earned an equally formidable reputation for litigation. Together, they transformed antitrust law and the international chemical industry.[6]

The opening salvo of the Justice Department's campaign against chemical cartels was fired in June, 1939, when a suit was filed against participants in the nitrogen agreement, including Du Pont, I.C.I., and Allied Chemical and Dye, among others. Particularly alarming to the British, not to mention Du Pont, was Arnold's readiness to drag foreign companies into court as "co-conspirators." In 1939 and again in 1942 Antitrust Division lawyers launched surprise "raids" on I.C.I.'s New York office, carrying off filing cabinets filled with documents that were later introduced into the cartel cases in court.

After lengthy negotiations, the nitrogen cartel case was settled in 1942 through a consent decree, the assumption being that the war had already disrupted the functioning of this arrangement, in which I. G. Farben had played an important part. Meanwhile, however, the Justice Department continued its systematic effort to ferret out arrangements deemed to be cartels. In 1942 Du Pont. I.C.I., and six other companies were charged with conspiring to monopolize the world dye industry; and in that same year Du Pont faced two separate suits focusing on its connections with Rohm and Haas in acrylic products and methyl methylacrylate. In 1943 a case was brought against National Lead Company for using the patent laws to suppress competition in the field of titanium pigments, and once again Du Pont and I.C.I. were haled into court for their participation in this extremely intricate system of patent exchanges. Du Pont people began keeping box scores of antitrust suits: one reckoning in 1955 had Du Pont winning eight out of eighteen suits, with eight settled by consent decree.[7]

Inevitably the 1929 and 1939 patent and process agreements between Du Pont and I.C.I. came under increasingly intense scrutiny. Up to the middle of 1942 most of the antitrust cases in chemicals had focused on arrangements involving the Germans, with an emphasis placed on the contribution of cartels to America's lack of preparedness for war. As one of Arnold's staff noted in a memorandum in September, 1942, however, this approach was "not suitable" for the Du Pont–I.C.I. agreements which was expected to be "the most significant single case" of the entire antitrust campaign in the chemical field. The two companies, aware of their vulnerability, had skillfully flanked the Justice Department by appealing to the U.S. Navy and War departments to intervene on their behalf, on the grounds that coping with a major lawsuit would interrupt their respective activities in Allied war production. This manuever postponed but could not block the eventual filing of a suit against Du Pont, I.C.I., and Remington Arms Co. in January, 1944.

Although Du Pont's lawyers had worked energetically throughout the interwar years to design patent exchange agreements that would be antitrust-proof, once the evil moment arrived, they and their corporate masters were prepared to jettison the agreements rather than undertake costly court battles. While the case was proceeding during 1945 and 1946 in the customary stately fashion of antitrust suits, Du Pont and I.C.I. began renegotiating their agreements in anticipation of what they presumed would be the ruling of the court. The nylon exchange was transformed into a conventional licensing contract, which seemed to benefit I.C.I. by extending its nonexclusive territory, but also resulted in the abrupt termination of new research information from Du Pont. This step was followed by a more revolutionary measure: the abandonment of the cross-licensing provisions of the patent and process agreement in June, 1948, which ended virtually all technical exchanges and "unceremoniously knocked away . . . the centerpiece of I.C.I.'s foreign policy . . . [and] the ideas by which, for nearly three generations, the chemical industry of the world had regulated its affairs."[8]

Du Pont had taken these unprecedented actions in part to persuade the Justice Department to accept a consent decree settlement. But the Antitrust Division, now headed by Tom Clark, remained unconvinced that the Du Pont–I.C.I. relationship had at last been obliterated. The I.C.I. case was brought to trial in April, 1950, and lasted nearly three months. After a year of leisurely cogitation federal district Judge Sylvester Ryan handed down an opinion and after another year of legal bickering a final judgment was passed, formally ending an alliance that in one form or another had endured for more than half a century.

Ryan's decree confirmed the termination of the various patent exchange agreements, but went further, ordering the dissolution of the joint ventures in Canada and South America, except for the Chilean explosives company, C.S.A.E., which was forbidden to regulate competition between the two major shareholders. Although Du Pont and I.C.I. had argued during the trial that the joint ventures were legitimate manufacturing operations, not a masquerade for a cartel arrangement, they complied with the dissolution order with little reluctance. In Argentina, Du Pont took over Ducilo, the rayon venture, and I.C.I. kept the rest, while in Brazil Du Pont retained the relatively small explosives operation and some miscellaneous fields. Canada's C.I.L. was the largest and the most complicated of the joint ventures to divide. Du Pont insisted on controlling the two most profitable of C.I.L.'s lines, nylon and cellophane, arguing that I.C.I. was not directly involved in either field (I.C.I.'s nylon undertaking was a joint venture with Courtaulds). I.C.I. took over the explosives, heavy chemicals, and Fabrikoid lines. The division was by no means a simple matter. In one case a line was drawn down the center of a C.I.L. factory that produced nylon and hydrogen peroxide. When I.C.I.'s representatives proposed to retain the name C.I.L. for their portion, Du Pont's Wendell Swint demanded $4 million for the concession; I.C.I.'s Peter Allen countered with an offer of $2 million. In the end, the decision was made by cutting a deck of cards, with Du Pont winning by drawing the high card.[9]

The alacrity with which the two companies complied with the court order reflected the changes that had occurred during and after World War II. By 1952 a new generation of executives was coming to power in both organizations, less committed to the traditions of cooperation, and intent on developing their new technologies without the inhibitions imposed by the patent and process agreements. Those agreements had already been virtually gutted in 1948, so that people on both sides became increasingly aware of the substantial differences in commercial strategies and management that distinguished I.C.I. from Du Pont and had always created tensions within the joint ventures. Finally, although neither Du Pont nor I.C.I. raised the subject in the 1950s, one of the major reasons for the "Grand Alliance" had been eliminated in the aftermath of the war as I. G. Farben fell victim to the American crusade against cartels and German militarism. The dissolution of the German chemical giant was probably as important as the I.C.I. case in reshaping the international chemical industry between 1945 and 1952.

The End of I. G. Farben

While the executives of I. G. Farben projected an image of confidence and arrogant certainty in their dealings with competitors and would-be cartel associates in the 1930s, privately they were obliged to recognize the precariousness of their situation. The Great Depression struck the German chemical giant at just the moment that I. G.'s financial resources were fully committed to a massive development program in synthetic fuel. Shortly before World War I, Friedrich Bergius, a B.A.S.F. chemist, had developed a method of producing gasoline from coal based on a high pressure synthesis process called hydrogenation that resembled the Haber-Bosch process for nitrogen extraction. During the 1920s Carl Bosch, anticipating oil shortages and rising prices, had promoted the synthetic fuel project. B.A.S.F.'s need for substantial capital investment to expand its hydrogenation works at Leuna had been a major incentive for the reorganization of I. G. Farben in 1926. The slump that began less than three years later, however, not only severely reduced sales in the traditional dye and fertilizer markets, but also undermined oil and gas prices, leaving I.G. with a "vested interest in a white elephant."

In desperation Bosch and his associates turned to the German government for aid. The Weimar regime imposed protective duties on oil imports to bolster I.G.'s hydrogenation investment, but these measures proved inadequate. The advent of the Nazis in 1932–33 offered some hope for further assistance, but also posed fresh problems for I.G. The synthetic fuel project meshed well with Hitler's notions about Germany's need for self-sufficiency, and he endorsed it warmly in conversations with I.G. officials in the fall of 1932 shortly before his elevation to chancellor of Germany. But Nazi anti-Semitism grated on Bosch, and Hitler's demands for "Aryanization" of German businesses led to friction between I.G. and the government in 1933. Meanwhile the company was provided with subsidies to develop synthetic fuels, but these funds did not benefit I.G. over the long run since all earnings above costs had to be turned over to the government. I.G. in fact never recovered the full cost of its hydrogenation investment between 1927 and 1945.[10]

Nevertheless, the links between I. G. Farben and the Third Reich became much stronger in the ensuing years as Bosch and Duisberg were succeeded by a new generation of executives more amenable to Nazi demands. Of particular significance was the relationship between I.G. and the German military fostered by Carl Krauch, who had headed the Leuna hydrogenation project in the 1920s, and Hermann Schmitz, who was chief managing director of the firm between 1934 and 1939 and succeeded Bosch as chairman of the board in 1940. Krauch cultivated relations with Hermann Goering, head of the German air force and after 1936 the dominant figure in economic preparations for war. By 1938 Krauch had become Goering's chief adviser on industrial mobilization, advocating measures that would commit Germany to a rapid buildup of its synthetic fuel and synthetic rubber capacity, expanding I.G.'s production, and guaranteeing the firm a central role in economic planning.[11]

As I.G.'s domestic operations became more closely integrated with the military preparations of the German government, its relations with foreign businesses through cartels and patent agreements shifted as well. During the 1920s the German company had fostered these arrangements to promote market stability and enable it to maintain a technological edge over potential competitors. The cartels now provided I.G. with some unanticipated opportunities. Through the patent and know-how exchanges, information could be acquired about technical processes potentially available to countries with whom Germany might soon be at war. The cartel system could also be exploited to procure vital raw materials for German industries or to obstruct military preparedness by Germany's rivals. At the same time, this exploitation of the cartels for military purposes could be invoked by I.G. to defend its connections with foreign firms against charges that periodically were raised by Nazis that I.G. placed profits over patriotism.

From this perspective, the most significant of I.G.'s array of cartel and patent exchange agreements was the link forged with the American oil behemoth Standard Oil (New Jersey) that dated back to the late 1920s. When B.A.S.F. had begun to move into the synthetic fuel field, it had naturally attracted the attention of Standard, a major supplier of oil to Germany, which had also begun investigating new methods of oil extraction from shale and coal during this period. In 1927 Standard and I.G. established a joint venture to introduce hydrogenation in the United States, restricting Standard's market in exchange for full technical disclosure by the Germans. Between 1927 and 1930 I.G. and Standard continued to negotiate on hydrogenation and related matters, culminating with the establishment of another joint undertaking, Jasco, Inc., that held patents to a range of coal and oil-based chemical processes developed by I.G. One of the most important of these processes was a synthetic rubber called buna, based on a combination of butadiene and styrene, that both the Germans and Americans believed could be produced more cheaply than the natural substance and would undercut the Anglo-Dutch cartel in that field.

I.G.'s foray into America via the Standard Oil connection was particularly irritating to Du Pont, which was already alarmed by the Germans' circumvention of the chemical tariff through General Dyestuffs and I.G. Chemical Co. Du Pont had no direct interest in hydrogenation at this point, but had begun work in the synthetic rubber field based on Carothers's research on polymers that would lead to the development of neoprene. Du Pont's suspicions were enhanced by the convoluted structure of patent agreements that crisscrossed the chemical and oil industries. I.C.I. was interested in hydrogenation and hoped to arrange a joint venture in the field with I.G. and Shell Oil. When I.C.I. negotiated an agreement with Standard Oil for access to the Standard-I.G. licenses in synthetic fuels in 1930, Du Pont raised vigorous objections based on its own agreements with I.C.I., pointing out that it would be possible for both Standard and I.G. to obtain information on Du Pont's research in synthetic rubber and in the high pressure synthesis area generally through I.C.I. After considerably bickering, Du Pont conceded I.C.I.'s right to proceed with agreements in the hydrogenation field with Shell, Standard, and I.G., but the episode damaged relations between the two companies to some extent and led Du Pont to take a greater interest in Standard Oil's activities in petrochemicals as well as its connections with the Germans.[12]

The collapse of oil prices after 1930 virtually brought to a standstill any further development of synthetic fuels outside Germany. Nevertheless, Standard Oil and I.G. maintained close ties, particularly in the area of synthetic rubber development. The Germans needed to produce abundant supplies of butadiene to reduce the costs of their buna process, and by 1930 had determined that this objective could best be achieved by extracting butadiene from acetylene derived from refined gas. I.G. turned over its know-how in the field to Jasco, and Standard Oil researchers began working on the process. Despite some initial problems, by the middle of the decade Standard Oil had developed methods of producing butadiene cheaply, thanks in large measure to their advances in the catalytic cracking of oil that also enabled the oil company to produce high octane gasoline for aviation fuel.

By this time I.G. had forged its links with the German military, which was pressuring the firm for a rapid buildup of synthetic fuel and synthetic rubber stockpiles. In 1937 Krauch and Schmitz procured a large supply of tetraethyl lead for the German Air Ministry from Standard Oil's affiliate, Ethyl Corporation. Shortly thereafter I.G. acquired through Jasco the know-how to Standard Oil's new process for synthetic rubber from oil so that it could move quickly to increase supplies for military preparations.

Du Pont learned of the Germans' interest in tetraethyl lead through its contacts with Ethyl as early as 1934 when I.G. was negotiating with that company to set up a joint venture in Germany. At that time Du Pont sent a sharp warning to Ethyl that its product could be used for aviation fuel and should not be transferred to the Germans for possible military purposes. Standard Oil took the precaution of clearing the proposed arrangement with the U.S. War Department before proceeding.[13]

While the War Department exhibited little immediate concern over I.G.'s efforts to acquire U.S. technology, Du Pont was less sanguine and began to take steps to disengage itself from the German firm without disrupting those arrangements that were of commercial benefit. In 1934 Du Pont sought to dispose of its shares in I.G. that it had acquired when the German chemical company had absorbed the explosives manufacturers D.A.G. and Köln-Rottweiler, in which Du Pont and I.C.I. held significant stock in 1925. Germany's exchange controls posed an obstacle to the transaction, but by 1940 Du Pont had rid itself of the shares, though at a loss of $1.2 million. Du Pont maintained its joint ventures with I.G. and continued technical discussions in a variety of fields, including synthetic rubber, but exercised a good deal more caution than Standard Oil in its dealings with its erstwhile foreign rival.[14]

By 1938 Standard Oil was also beginning to have second thoughts about its ties to I.G. Despite its gestures of goodwill in transferring data on tetraethyl lead and synthetic rubber, Standard had received little reciprocal information on I.G.'s progress with buna, largely because the German government had vetoed such a transfer. I.G. had also refused to allow Standard to license Jasco's patents to other American firms. In August, 1939, with the outbreak of war imminent in Europe, Standard Oil hurriedly arranged to purchase I.G.'s shares in Jasco to prevent confiscation of the joint venture should the U.S. enter the conflict. A month later Standard's Frank Howard met with I.G. officials at the Hague and arranged at last for the transfer of buna patents, but the Germans declined to transmit information on know-how, a typical I.G. tactic that Standard would later regret having conceded.

Despite its efforts to establish an arms-length relationship with the German firm for the duration of the war, Standard Oil remained vulnerable to criticism that did not take long to appear. Both the U.S. Justice Department and the Treasury began investigating I.G.'s American contacts in 1940, and Jasco was a particularly juicy item. By the end of 1941 Thurman Arnold was pressing for a criminal indictment of Standard Oil, which responded first by persuading the War Department to demand postponement of the case, then by agreeing to plead *nolo contendere* to the criminal charges and accepting a consent decree in a civil suit filed at the same time, under which Standard would license all the patents it had acquired from I.G. on a royalty-free basis.

Arnold was not finished with Standard Oil. Worried that intervention by the War and Navy departments on behalf of corporations such as Standard and Du Pont would slow the momentum of his antitrust drive, Arnold sought to dramatize the malign effects of international cartel agreements on America's war preparedness. Appearing before Senator Harry Truman's special committee on war mobilization in March, 1942, Arnold charged that current shortages in rubber supplies following the Japanese conquest of Southeast Asia were largely attributable to Standard Oil, which had delayed development of synthetic rubber at the behest of its cartel partner, I. G. Farben. Standard Oil officials denied the allegations, arguing that they had acquired more valuable information, including the buna patents, than they had given away.[15]

Although Standard Oil survived its "ordeal of 1942" with little lasting damage, Arnold's revelations generated a widespread public concern over American links with German businesses, particularly in the chemical industry. Du Pont, despite its cautious disengagement from I. G. Farben culminating with the suspension of patent exchange agreements in nylon, styrene, and MVA in April, 1941, did not escape unscathed. Justice Department interest focused on the 1936 agreement made by Du Pont with Rohm and Haas in acrylic plastics and laminated glass. Since the latter material was used in airplanes, Du Pont could be said to have turned over a field of potential military importance to a German firm, although the connection was more indirect than in the Standard Oil–I.G. situation. Du Pont had canceled the agreement in February, 1941, and substituted an arrangement for an exchange of licenses without restrictions, but Justice wanted the process on laminated glass released on a royalty-free basis. The military departments intervened at this point, arguing that the issue should be postponed until after the war and that access to the process should continue to be restricted. In 1945, when the case reemerged, criminal charges against Du Pont were dismissed and a consent decree settled the civil charges.[16]

While the U.S. government sought to chasten American firms for their pre-war dealings with German counterparts, it had more drastic designs in mind for I. G. Farben. Schmitz and his associates were aware, long before the outbreak of war, that their investments in the United States and perhaps elsewhere in the Western hemisphere were at risk if America entered the conflict, and they cast about for ways to prevent a repetition of the experience of confiscation in World War I. To that end I. G. Chemie, which had initially been established as a Swiss tax haven for their foreign earnings, presented new possibilities. Its American affiliate, I.G. Chemical Co., already had a board composed of reputable American businessmen, including Walter Teagle of Standard Oil and Edsel Ford. In 1936 Schmitz and Bosch resigned from the board of American I.G., and in 1939 the company changed its name to General Aniline and Film Co. Through a convoluted series of stock manipulations, I. G. Chemie's German link was nominally severed, so that General Aniline could be presented to American authorities as a Swiss subsidiary.

These efforts at camouflage proved futile, however, as Justice, Treasury, and the Securities and Exchange Commission in the U.S. were certain that General Aniline and its affiliate, General Dyestuffs, remained under German control. In 1940 the U.S. government blocked the transfer of all foreign assets and earnings, and in December, 1941, shortly after Pearl Harbor, Attorney General Francis Biddle pushed a revision to the Trading With the Enemy Act through Congress that authorized the government to seize the assets of any foreign firm that might be providing aid to the Germans. Four months later General Aniline and General Dyestuffs were "vested" by the U.S. Alien Property Custodian, which also took over Jasco over the protests of Standard Oil.

After the war the fate of these enterprises became a matter of extensive litigation. In 1945 Standard Oil sued the government for recovery of Jasco, but had to settle for a return only of its shares in the joint venture and patents held before 1938, while the government retained control of I.G.'s shares and the buna patents. In 1953 Standard finally reacquired the rest of Jasco by bidding $1.2 million for the shares at a public auction staged by the Alien Property Custodian. The disposition of General Aniline was more complicated. After the war I. G. Chemie, reorganized as "Interhandel" A.G. and apparently cleansed of German ties, sued for recovery of 89 percent of General

Aniline at an estimated market value of $100 million. The case dragged on for years through the U.S. courts and the International Court at the Hague, and encompassed much diplomatic haggling as well among German, Swiss, and American governments. Finally in 1965 U.S. Attorney General Robert Kennedy arranged for the public sale of General Aniline for $329 million, with the proceeds divided between the American government and Interhandel at "the largest competitive auction in Wall Street history."[17]

By the end of World War II I. G. Farben had become for the American public the virtual embodiment of the sinister connection between German big business and the Third Reich. As the Allied armies moved into the German-occupied regions of Europe, the extent of I.G.'s involvement in the Nazi slave labor and extermination camps was revealed. In these circumstances unprecedented proposals for the breakup of the German chemical giant found widespread support, even among conventionally conservative business and political figures in the United States such as Bernard Baruch and Dean Acheson.

A variety of motives lay behind these demands. There was genuine outrage among the Western Allies over the participation of I.G. in Nazi war crimes and I.G.'s crude takeovers of chemical concerns in occupied territories. In 1947 charges were brought against twenty of I.G.'s executives, including Schmitz and Krauch, at the Nuremberg Tribunal. After almost a year of hearings the judges ruled that a number of these men were guilty of involvement in mass murder, slavery, and plunder, although they rejected additional charges that I.G. had connived with the Nazi regime in planning the war. The American prosecutors regarded the sentences as excessively mild, ranging from one to eight years' imprisonment. Nevertheless, the Nuremberg trial of I.G.'s chiefs represented a unique proceeding against executives usually well insulated from the consequences of their actions.[18]

More mundane considerations of policy animated other measures taken against I. G. Farben. In 1945 American and British oilmen suddenly appeared in Germany, replete with temporary military titles, as part of a "Technical Oil Mission" that systematically rummaged through I.G.'s hydrogenation plants and carted off masses of documents from the company's headquarters at Frankfort. I.G. chemists involved in synthetic fuel and synthetic rubber development were interviewed and some were recruited by American firms. The clear intent of this operation was to procure German technical secrets for the benefit of the victors, although by the early 1950s the oil industry, awash in a flood of inexpensive oil from Venezuela and the Middle East, lost interest in the subject of synthetic fuels.[19]

During this same period Allied authorities in occupied Germany began to "deconcentrate" I.G. For the Russians these measures were treated as simply reparations collected under another name, and they set about dismantling

I.G.'s plants in their own zone. The Americans and British proclaimed a more elaborate rationale for their proposed actions. During the war Arnold and his supporters in Congress had argued that unless I.G. Farben was broken up the international chemical cartels would quickly reemerge, probably with the connivance of American firms such as Standard Oil. Germanophobes such as Treasury Secretary Henry Morgenthau, Jr., warned that I.G., if left intact, would provide a revived Germany with the wherewithal for yet another round of rearmament and aggressive war. The British government did not necessarily accept these arguments but endorsed deconcentration on the reasonable assumption that anything that weakened I.G. might indirectly benefit their own chemical export industry.

The Americans were naturally most vigorous in promoting deconcentration. In 1947, after enacting a general antitrust law, the U.S. occupation regime under General Lucius Clay announced that I. G. Farben's holdings within their zone had been divided into forty-seven "independent units" which were to be barred from collaborating with one another. Within a year, however, the momentum of deconcentration slowed as the U.S. government began to emphasize the need to rebuild the German economy to meet the challenges of the Cold War. In 1950 the Anglo-American Allied High Commission in Germany, which had been established following the breakdown of relations with the Soviets over the reunification of Germany, issued a new law concerning I. G. Farben that would ensure the organization of "economically sound and independent companies" in the German chemical industry. Former shareholders in I.G. lobbied effectively with the new West German government in Bonn to block any widespread dispersal of the assets. In 1953 the High Commission finally approved a plan that would resurrect the three largest constituent elements of I.G.—Bayer, Hoechst, and B.A.S.F.—which would assume control of most of the assets of the consolidated firm.[20]

The onset of the Cold War had helped the German chemical producers avoid the full impact of America's postwar plans for deconcentration, but the loss of foreign affiliates and the general dispersion of many of the German patents throughout the industry, together with the physical destruction of the I.G. plants in Germany—the synthetic fuel and buna plants had been major targets of Allied bombing in 1944–45—and the disgrace of former I.G. leaders constituted substantial blows to the prewar position of the German chemical manufacturers and temporarily removed them as major actors on the international scene. With the concurrent breakup of the Anglo-American alliance, the transformation of the postwar chemical industry was complete. The new system that would ultimately replace the world of cartels and patent exchanges in chemicals would take some time to become clear.

CHAPTER THIRTEEN

Du Pont Goes Abroad

While World War II and the American campaign against cartels disrupted arrangements in the international chemical industry, the domestic U.S. market was being transformed as well. Total sales in American chemicals more than tripled in dollar value between 1939 and 1949, and total production quadrupled. Chemical stocks acquired an unprecedented glamour on the stock exchanges, trailing only oil and electric utilities in trading volume. While Du Pont's polymerized fibers caught the headlines, there was significant growth for a vast range of chemical products. DDT, first used extensively by the U.S. Army in Italy and the Pacific during the war, was introduced to the public in 1946, ironically (in retrospect) hailed as "the atomic bomb of the insect world," and paved the way for a variety of insecticides and herbicides. The war also promoted the rapid growth of new pharmaceuticals, notably penicillin and sulfa drugs, first produced on a commercial basis in 1942 for military operations in the malarial South Pacific. New forms of polymerized plastics, such as polyvinylchloride and polystyrene-based materials, developed principally by I. G. Farben before the war, were introduced into the American market by Union Carbide and Du Pont. Synthetic rubber, a major achievement of wartime research and development, increased output from 20,000 tons in 1942 to over 700,000 tons by 1945, and rapidly displaced the natural product as more versatile forms were discovered.

Growth and innovation were accompanied by structural changes in the postwar chemical industry. Military demand for basic chemicals during the war had created shortages in the civilian market, leading some firms in fields dependent on processed materials such as glass, paint, film, and foods to integrate back to ensure supplies, as Du Pont itself had done in World War I. The prospect of rapid growth attracted outsiders, particularly oil companies, into the chemical fields. This proliferation of part-time chemical producers would create problems later for the major firms, but in 1945–57 the market was large enough to absorb the newcomers with room to spare.

The "Big Three" of the prewar domestic chemical industry—Du Pont, Union Carbide, and Allied Chemical—maintained their position, but there were portents of change. Allied Chemical had benefitted from wartime demand for synthetic ammonia and dye intermediates, but the legacy of Weber's financial conservatism and indifference to technological innovation and diversification was beginning to have its effect on the firm. In 1939 Allied's net sales were equal to those of Union Carbide, and 56 percent of Du Pont's total. A decade later, Allied's sales were at 60 percent of Union Carbide, and little more than one third those of Du Pont. The sheer size of the company helped prop up Allied's earnings and market position, but it was soon to be challenged by more aggressive and innovative firms.

One of the major beneficiaries of growth in the industry during the war was Dow Chemical, whose sales skyrocketed over 500 percent in 1939–49, and tripled again in the following decade. The company's founder, Herbert H. Dow, had studied chemistry at the Case School of Applied Science in Cleveland, which in the late nineteenth century was the oil-refining capital of the country, and had become interested in developing a process to extract bromine, used in pharmaceuticals, from the brine residue of natural gas wells. After several years of experimenting, Dow perfected an efficient process based on electrolysis, and in 1897 moved his operation to Midland, Michigan, where there were abundant brine deposits. Subsequently he developed a method of extracting chlorine from brine and began producing the bleaching agent on a commercial basis, soon emerging as the major American competitor in that field with the British firm, United Alkali. During World War I, Dow jumped on the dyestuff bandwagon, developing one of the first domestic synthetic indigo dyes in 1916, and producing chlorine gases for the military that led the company into research on the uses of ethylene. In the following decade, Dow initiated a project to extract bromine from sea water that attracted the interest of G.M. researchers Kettering and Midgley, who were developing their bromine-based antiknock fuel. Through this connection, Dow procured a supply contract for ethylene dibromide with Ethyl Gasoline Company.

Dow's growth and incipient entry into directly competitive fields did not go unnoticed by Du Pont. In the midst of its 1927 buying spree, Du Pont approached Dow with a tentative merger bid, but the aging inventor-entrepreneur, who retained a controlling minority interest in the $20 million company, rebuffed the offer. Two years later Herbert Dow died, succeeded by his son

Willard, who continued the founder's emphasis on diversification through research, moving into styrene-based plastics, pesticides, and petrochemicals.

One product of special significance in Dow's future was magnesium. During World War I Herbert Dow had taken an interest in this extremely light alloy metal similar to aluminium, anticipating its use in airplane frames. The prospect of deriving magnesium from brine helped propel Dow into his experiments with sea-water extraction. As one of the few companies involved in this field in the United States, Dow benefitted from the 1922 tariff rates on magnesium imports, but development lagged as the company encountered difficulties in alloying the metal and locating immediately exploitable markets. In 1927, however, Dow negotiated an arrangement with American Magnesium Corporation, an Alcoa subsidiary, that left it the sole producer of the metal in the United States, and provided Dow with valuable patents.

Over the next seven years Dow was involved in marathon talks with I. G. Farben as the German company, which had pioneered in the development of magnesium since 1913, threatened to enter the U.S. market. In 1934 Dow, Alcoa, and I.G. arranged a patent-exchange agreement similar to the Du Pont–I.C.I. agreement that left Dow with its magnesium monopoly in the United States, with foreign sales only to I.G. This agreement ran Dow afoul of the U.S. government when Thurman Arnold mounted his anticartel campaign. The company was fined after pleading *nolo contendere* in 1941, and licensed its magnesium patents on a royalty-free basis to other producers at the government's behest when war production began. But Willard Dow vigorously defended his firm throughout the proceedings, arguing that Dow's "monopoly" had gained her little since the company had been producing at a loss for all but three years between 1918 and 1939, and that its main purpose in negotiating the 1934 deal with I. G. Farben was to enable Dow to expand its sales volume and keep the magnesium operation afloat.

Wartime demand for magnesium for aircraft and other uses was a major contributor to Dow's expansion after 1941, and its work in the polystyrene field provided Dow with an important role in the government's synthetic rubber program as well as a foothold abroad, when Dow assisted the Canadians in establishing a $50 million rubber plant at Sarnia, Ontario. With assets of $200 million and sales running at $170 million at the end of the war, Dow was still relatively small, but the time was fast approaching when Du Pont people would regard Dow, not Allied, as their principal rival on the American chemical scene.[1]

Another company that was attracting the interest of postwar investors was Monsanto Chemical Co. of St. Louis, Missouri. Established in 1901 by John

F. Queeny to manufacture saccharine, it gradually expanded, largely through acquisitions in the 1930s, into dye intermediates, heavy chemicals, and plastics, primarily as an industrial supplier. Monsanto's growth in the war was not as dramatic as Dow's, but it picked up momentum in the postwar decade. By 1955 Monsanto's sales volume and share price was equal to those of the Michigan firm.

Monsanto's expansion was due in large measure to its entry into the synthetic fibers field. In 1949 Monsanto joined with American Viscose, the formerly British-controlled rayon company, to establish Chemstrand Corporation. Monsanto had begun experimenting with acrylic fibers in 1942, and relied on Viscose with its links to the textile industry to handle marketing and distribution. At the outset the joint venture ran into trouble, since Monsanto's new fiber was inferior in quality to Du Pont's orlon acrylic, but the company became the inadvertent beneficiary of Du Pont's antitrust difficulties. In 1950 Greenewalt approached Monsanto with a proposal to license nylon to Chemstrand in order to counter the charges of a nylon monopoly advanced in the I.C.I. suit, and also to establish a basis for determining the commercial value of the patent should Du Pont be required to license nylon on a nonexclusive basis. In addition to the patents and know-how, Du Pont helped Chemstrand set up a nylon plant at Pensacola, Florida, for $120 million and a ten-year royalty arrangement. By 1953 Chemstrand was producing over 50 million pounds of nylon, expanding later to over 200 million pounds per year.[2]

In effect, Du Pont created its own competition in nylon, albeit under some duress. By the end of the 1950s the nylon success story had attracted a variety of newcomers into the polymer plastics and fiber fields. I.C.I. had developed polythene, which was licensed and improved upon by Du Pont. Polyester and spandex fibers were introduced commercially in the middle of the decade, and by 1960 production of fibers and plastics was beginning to outpace sales, with the threat of serious overcapacity looming in the near future.

Du Pont was large enough and its product lines so diversified that even the erosion of the synthetic fibers market did not pose an immediate danger. But all the chemical manufacturers faced difficulties by the end of the 1950s. While total sales by the major firms continued to grow, a lengthy steel strike in 1956 and a sharp economic slump the following year cut into net earnings as manufacturers reduced prices on nylon, styrene, and ammonia in order to maintain volume. Du Pont's net profits on sales leveled off at about 22 percent after peaking in 1955, and net return on investment fell from 26 to 20 percent as the Wilmington company, like other American chemical producers, was obliged to maintain plant and research expenditures in order to hold its competitive position. The post 1957 decline in share prices hit the high-riding chemical makers particularly hard, with Du Pont down 22 percent from its peak value in the preceding decade. Dow went down 28 percent and Allied and Monsanto by more than one third, compared to an average 10 percent decline for other major industrial stocks.[3]

These discouraging trends engendered much soul searching in the boardrooms of the chemical majors and would shortly lead to the adoption of new strategies. Du Pont's problems with declining earnings and saturated markets—problems that it shared with others—were aggravated, however, by continuing confrontations with the Justice Department that culminated in the U.S. Supreme Court's order that Du Pont sunder its connections with General Motors. While the G.M. divestiture would not fundamentally alter the structure and operations of Du Pont, it did mark a turning point in the company's history and contributed to the atmosphere of crisis and change in direction that would shape Du Pont for the next twenty years.

The General Motors Case

Of the twenty-odd antitrust suits launched against Du Pont during and after World War II, two were remarkable for their duration and the public attention they received: one was the "Cellophane" case, focusing on Du Pont's alleged monopoly in this field and its relations with Sylvania Corporation; the other was the "General Motors" case, involving Du Pont's stockholdings in the auto firm and its alleged abuse of that position in supplying parts and materials. Both cases began in the mid-1940s and slowly wound through the courts for more than a decade. Both cases were contested by Du Pont on the legal and public relations fronts. The final results, however, were markedly different.

The Cellophane case was initiated in 1947 as part of Assistant Attorney General Wendell Berge's renewal of Thurman Arnold's efforts to promote competition in postwar markets. Du Pont's patent to moisture-proof cellophane had recently expired and Dow had just begun to market a new packaging material called Saran Wrap, but the Wilmington company was anticipating a major expansion of its cellophane capacity, since production had been curtailed during the war and the domestic market potential was substantial. Shortly after the civil suit was filed, Du Pont canceled its expansion plans and offered to license its know-how to firms whose previous requests for licensing had been refused. Olin Industries accepted the offer which, unlike the earlier arrangement with Sylvania, did not impose any restrictions on output. Du Pont also renegotiated its license with Sylvania and could now assert that there were three genuinely competing firms in the field.

Nevertheless, the Justice Department continued the suit. Antitrust lawyers were bolstered by the precedent established in the Alcoa case in 1945, where the court had determined that "market control" defined as "the power to control prices or exclude competition" was sufficient to prove a violation of the Sherman Act even where the dominant firm had not conspired with others to achieve its position or had not abused its monopoly through high prices. In the case of Du Pont's cellophane, as in the Alcoa situation, market prices had declined steadily since 1930 while earnings rose through high volume production. In 1953, however, Judge Paul Leahy of the federal district court in Delaware concluded that the government had failed to prove its case, arguing that Du Pont held only 67 percent of market capacity and had achieved its position earlier through its legitimate control of the cellophane patents.

The case was appealed to the U.S. Supreme Court which in 1956 upheld Leahy's judgment by a four to three vote. The main issue in contention was the definition of the cellophane market. Du Pont had argued, and Leahy agreed, that the market encompassed all forms of flexible packaging materials that were "reasonably interchangeable" with cellophane. The Supreme Court justices devoted an extraordinary amount of attention to this question, "including a trip to a packaging exhibition and a local supermarket to view the various means of merchandising potato chips and frozen foods." One peculiar aspect of this case, in view of the Justice Department's previous forays against Du Pont's foreign entanglements in the National Lead and I.C.I. cases, was the failure of the government attorneys to emphasize Du Pont's patent exchange agreement with the French. Judge Leahy had dismissed this matter on the ground that the market-sharing aspects of the Du Pont–Sylvania–La Cellophane agreement "were ancillary to the legitimate license of trade secrets"; but it had been on the basis of agreements quite similar to these that Du Pont had been judged in violation of the Sherman Act in the I.C.I. case.[4]

The favorable (for Du Pont) outcome of the Cellophane case, however, was followed in less than a year by the devastating, and generally unexpected decision of the Supreme Court in the General Motors case. This case too had been initiated almost a decade earlier but for more overtly political reasons than in the case of cellophane. In the summer of 1948 the Antitrust Division had begun filing a series of suits against big business, including Du Pont and General Motors. Anticipating, inaccurately, that Harry Truman would be defeated in the November elections, the Justice Department hoped to appeal to organized labor by taking on companies with a strong antilabor image and to reap political benefits later should the incoming Republican regime drop the cases. Although the Democrats remained in power and the grand jury declined to indict Du Pont and G.M., the Justice Department pursued the matter, filing a civil suit in federal district court in Chicago in 1949.[5]

The G.M. suit did not come as a complete surprise to Du Pont. During the 1920s, when the company was under pressure from Justice over its acquisition of U.S. Steel stock, the G.M. connection had also come under scrutiny, although the government took no further steps at that time. When antitrust prosecutions began to proliferate after World War II, the Du Pont–G.M. link provided a logical target. In 1946 Du Pont's president Walter Carpenter requested an analysis by the financial and legal departments of the possible tax consequences of a major distribution of the company's G.M. holdings to Du Pont shareholders, and in January, 1947, nine months before the company learned of the Justice Department's investigation, Carpenter proposed to the board that Du Pont issue preferred stock convertible to G.M. shares as a first step toward disengagement of the two firms. At this point, however, Pierre du Pont, still chairman of the board at seventy-seven, exercised his rarely invoked veto power. Pierre objected to this "change in Du Pont's policy of nearly thirty years standing which has been to retain an investment that has proven very valuable," adding the somewhat improbable observation that a Du Pont divestiture would irritate G.M. management.[6]

In 1948, after the antitrust suit was filed, Du Pont retained Hugh Cox, former U.S. solicitor general under President Roosevelt, to advise it on the case. Cox recommended that Du Pont sell its G.M. stock and avoid a court battle that would reap the company a harvest of bad publicity. Again Pierre refused to consider the option, and his view was supported by the incoming president, Crawford Greenewalt. Up to this point Du Pont, like other big business targets of antitrust, preferred to settle these cases where possible by pleading *nolo contendere* and negotiating with the Justice Department; but both sides were now committed to an extended contest. Cox assembled a force of more than thirty lawyers to oppose Justice's attorney Willis Hotchkiss, whose staff was poring over thousands of subpoenaed documents relating to Du Pont, G.M., and affiliated enterprises, including Ethyl Gasoline Co., Kinetic Chemicals, U.S. Rubber, and Bendix Aviation.

The basic argument of the antitrusters was that the 23 percent equity held by Du Pont in G.M., combined with the shares held by the Du Pont family members through two holding companies, Christiana Securities and Delaware Realty Co., created a situation of excessive concentration in both industries. Drawing on the arguments made by Raskob when Du Pont was considering the G.M. investment in 1919, Hotchkiss maintained that Du Pont intended from the outset to use its equity position to force G.M. to give the chemical firm preferential treatment as a supplier of automobile paints, fabrics, and associated materials, extended later to include tires produced by U.S. Rubber, tetraethyl lead produced by Du Pont for Ethyl, and freon produced by Du Pont for Kinetic Chemicals. These arrangements not only squeezed out competing suppliers of these products, but also resulted in lower cost sales by Du Pont to G.M., enabling the auto firm to undersell competitors in its own market. In presenting this argument, the government focused primarily on evidence of conspiratorial arrangements to block competition in violation of the Sherman Act, adding almost as an afterthought the charge that Du Pont's links to G.M. violated section 7 of the Clayton Act of 1914 that barred acquisition of stock by one company in another if the result was to "tend to create a monopoly in any line of commerce."[7]

Du Pont's response came both in the court and in the public media. Cox's staff assembled material to demonstrate that Du Pont's position as preferred supplier to G.M. was the result of its superior products, attested to by sales to other auto makers; and that G.M.'s managers had frequently resisted Du Pont pressures, rejecting proposals for "reciprocity" in purchasing in the 1920s, and following a consistent policy of maintaining several suppliers of parts and materials to ensure cost controls. Meanwhile, Du Pont's public relations chief Harold Brayman, an experienced journalist, inaugurated a new policy of open relations with the press. Where Walter Carpenter had recoiled from public interviews, Greenewalt made himself readily available and encouraged journalist's surveys of the wonders of Du Pont's technical achievements and its contribution to the nation's growth, prosperity, and security, a message that found a ready audience in Cold War America. When the Justice Department charged that Du Pont family members were using trust funds for their children to extend their control of G.M., Brayman gleefully released photographs of an infant Du Pont identified as a "hardened conspirator."[8]

In December, 1954, Judge Walter LaBuy ordered dismissal of the case, asserting that the government had failed to prove its conspiracy charges and that neither Du Pont nor G.M. exercised a monopoly in their respective industries as a result of their connection.[9] The Justice Department appealed the decision, but Du Pont's lawyers felt reasonably sure that the company had little to fear, since LaBuy's decision had focused on the inability of the government to show a factual basis of its claims. The legal issues involved appeared to be settled.

The Supreme Court, however, took a different view. Three of the Justices disqualified themselves from the appeal, two because of prior involvement with Du Pont on the case. Four of the remaining six voted to overrule Judge LaBuy, drawing on the Clayton Act provision on stock acquisitions, and asserting that the act as it applied in 1949 restricted vertical integration as well as horizontal combinations among firms in the same field.[10]

The Justice Department moved quickly to exploit its unanticipated victory, demanding that Du Pont dispose of its 63 million G.M. shares by sale, giving G.M. the first option to buy the stock and prohibiting Du Pont family members and their holding companies from buying any of the divested stock. In addition, Du Pont was to be obliged to abandon the tetraethyl lead field and sell its interest in Kinetic Chemicals. Du Pont contested these proposals, arguing that dispersal of such a huge block of shares even over a fairly long time period would depress the value of the stock by as much as 25 percent, thus hurting all G.M. investors. The Du Pont counterproposal was for a distribution of the company's shares in G.M. to Du Pont shareholders, on the basis of 1.38 G.M. shares for each Du Pont share; but even in this situation individual investors would suffer from the "confiscatory" tax rate that would be imposed, estimated to total $778 million. Russell Harrington, commissioner of the Internal Revenue Service, confirmed this pessimistic view in 1958, maintaining that the distributed stock would be regarded as dividends to Du Pont shareholders and thus subject to taxation, rather than as capital gains.

Judge LaBuy, to whom the case had been remanded, recommended after almost two years of hearings that Du Pont should "pass through" to its shareholders only those G.M. shares that carried voting rights in the automobile company, and that Delaware Realty and Christiana should be enjoined from exercising voting rights with the twenty-one million shares the two trust companies held in G.M. He believed this arrangement would eliminate the issue of Du Pont control of G.M. while protecting Du Pont shareholders from "harsh and punitive" taxation and a rapid decline in the market value of their shares.[11]

The Justice Department remained dissatisfied with the LaBuy proposal. Possibly influenced by the approaching 1960 presidential election, the Antitrust Division appealed again to the Supreme Court, demanding a complete divestiture of all G.M. stock by Du Pont, tempering its position by proposing a ten-year period of distribution to alleviate the tax problem. Again the Supreme Court by a four to three vote overturned LaBuy's "pass through" plan, arguing that "Common sense tells us that . . . there can be little assurance of the dissolution of intercorporate community of interest which we found to violate the law. . . . The Du Pont shareholders will ipso factor also be General Motors voters. It will be in their interest to vote in such a way as to induce G.M. to favor Du Pont. . . ."[12]

Greenewalt had begun lobbying in 1958 for congressional action to nullify the I.R.S. ruling. Delaware Senator Allen Frear introduced a bill that would allow the G.M. shares distributed to Du Pont stockholders to be treated as capital gains until they were resold. After the Supreme Court decision in May, 1961, the bill reemerged with the support of the newly installed Kennedy administration. Following its enactment in 1962 both sides quickly concluded an arrangement with LaBuy for the sale of the entire block of G.M. shares by Du Pont, Christiana, and Delaware Realty within three years. In 1965, forty-nine years after Pierre Du Pont's first investment in Durant's company, the links between the largest chemical manufacturer and the largest auto maker in the United States were finally terminated.[13]

Despite the complaints from both Wilmington and Detroit about the consequences of the divestiture, neither Du Pont nor G.M. was damaged irretrievably, and the outcome of the case was not wholly unwelcome. Market prices on both stocks fell sharply in the immediate aftermath of the Supreme Court actions, but leveled off as the economy began to improve in 1963. Since Du Pont had always segregated its G.M. dividends from earnings from its operations investment, the divestiture did not disrupt the company's financial structure. Earnings per share for Du Pont's own sales rose from $5.72 to $8.63 during 1963–65, although total earnings per share fell by $1.42 in this period due to the loss of G.M. dividends, and the aggregate value of the G.M. stock sold in three installments over this period came to $4.5 billion. For G.M. the separation had only short-term financial repercussions, but the effect on management was significant. Even before the Supreme Court decision in 1957 there had been signs of friction between the auto firm's executive committee, with its preponderance of production-minded technical men, and the New York-based finance committee, which included substantial Du Pont representation. By the end of the decade the habitually cautious Du Pont men were trying to rein in G.M.'s expansion plans, and the divestiture ensured the passage of operational control to the auto makers.[14]

Du Pont's loss, however, was more than superficial. Earnings from its G.M. investment had cushioned the company against hard times in the past, since these dividends were passed on to Du Pont shareholders, freeing more of earnings from operations for reinvestment. The full effect of the continuing cost-price squeeze on net operating earnings became more apparent with the progressive removal of G.M.'s one-third contribution to Du Pont's total profits.

The company had anticipated this situation since 1957, and had initiated measures to overcome the problem over the long run. Research and investment in newer areas such as high-resistant metals, and development of new products (including the ill-fated Corfam), was accompanied by disengagement from older lines such as rayon, which was abandoned in 1963, and dyestuffs, which were scaled down over the next twenty years and finally ended in the early 1980s. Of all Du Pont's strategies for recovery, however, probably the most significant in terms of the future direction of the firm was the decision to enter foreign markets both through exports and substantial direct investment abroad. Du Pont was by no means alone in taking this step. Other U.S. chemical firms, notably Dow and Monsanto, began to move abroad in the late 1950s, as did numerous other manufacturers whose previous foreign experience had been even more marginal than Du Pont's. Rising competition and labor costs at home and the lure of the Common Market in Europe and emerging nations in Latin America and Asia contributed to a major shift in American capital investment in the years 1957–67. At the same time, the particular concerns of Du Pont and changes in its outlook and management were equally important factors in the decision to go abroad.

The Road to Foreign Investment, 1945–62

In 1939 Du Pont's export sales grossed $14.8 million, about 5 percent of total sales. More than one third of its exports went to Latin America, and three manufacturing departments—Organic Chemicals, Fabrics and Finishes, and Explosives—accounted for 40 percent of foreign sales. These three departments and Du Pont's arms subsidiary, Remington, had established export division with branch sales offices abroad. In addition, the company had $20.6 million invested in foreign companies, about 4 percent of total investment (including the equity in G.M.). Aside from some minor subsidiaries such as an explosives manufacturer in Mexico, the foreign affiliates fell into two categories: joint ventures with I.C.I., including C.I.L. and the Duperial ventures in Latin America; and the Duco enterprises in Europe, in which Du Pont held minority shares and licensed the patents and know-how. Foreign direct investment was thus largely a byproduct of Du Pont's international patent agreements, and exports represented an insignificant proportion of the company's general operations.[15]

By the end of 1944 Du Pont exports had increased to $39.4 million, about 6 percent of total sales, but Lend-lease sales, dependent on U.S. wartime spending, comprised 38 percent of this total. Nevertheless, Du Pont's Foreign Relations Department believed that sales could be maintained at this level after the war, and that its Foreign Trade Development Division, which had remained virtually dormant since its establishment in 1929, should be assigned responsibility for coordinating the new export drive. In addition, the department lobbied for the expansion of local manufacturing subsidiaries, particularly in Latin America where Ducilo in Argentina and Duperial in Brazil were accumulating reserves that could be reinvested without drawing on the parent firm, and where postwar industrial development looked promising, with American aid flowing into the region.[16]

None of these initiatives proceeded very far. Du Pont's export sales did increase in dollar volume, to $62 million in 1949, more than doubling to $144 million over the next decade, but exports as a proportion of total sales remained fairly constant, hovering at 6 to 8 percent up to 1957. The manufacturing departments resisted coordination of exports and the two most active ones, Organic Chemicals and Fabrics and Finishes, expanded their own distribution and sales organizations in Europe. Furthermore, the manufacturing departments regarded Du Pont's foreign subsidiaries as competitors even after the joint ventures in Latin America and Canada were divided. Under Du Pont's decentralized organization, a department sales manager "didn't care who he lost business to, just that he lost business."[17]

The attitude of Du Pont's Executive Committee toward foreign direct investment was equally discouraging. Proposals to establish a heavy chemicals plant in Brazil and a rayon plant in Mexico were vetoed during World War II. After the war Du Pont began to reduce its foreign commitments, particularly with the Duco companies in France and Italy, and rebuffed the overtures of Solvay of Belgium to extend their European joint ventures. During the 1930s Du Pont had come to regard the Duco companies as increasingly irksome entanglements. The French connection was particularly painful. Du Pont held a 35 percent share in Societe Francaise Duco and was entitled to more than half the profits, but as the depression worsened, losses mounted and the company remained afloat only through the injection of loans from Du Pont's partner in the venture, the French Nobel company. Du Pont's representatives on the board repeatedly recommended changes in policy and management, but met only defiance from S.F.D.'s chairman, a "charming but incompetent businessman." By 1935 Du Pont was prepared to pull out of the company, but balked at the terms offered by the French. The war interrupted Du Pont's contacts with the Duco firms: the German Duco affiliate, a joint venture with Schering Kahlbaum A.G., ceased paying dividends to Du Pont after the Nazi regime imposed exchange controls in 1934, and its main plant was later destroyed by the Russians. In 1946 the Wilmington company decided to pull out of these activities that seemed to be permanent drains on Du Pont's time and resources. S.F.D. was hastily sold off to the French Nobel, and the Italian Duco was acquired by Montecatini. The whole Duco episode seemed to demonstrate the futility of entering foreign markets through joint ventures with partners less amenable or vigorous than I.C.I.[18]

Not all the members of Du Pont's executive committee were hostile toward international investment. Walter Carpenter, Jr., who had begun his career with Du Pont in its Chilean subsidiary before World War I, supported the postwar export drive and was inclined to endorse the arguments of the Foreign Relations Department that some foreign manufacturing capacity would ultimately have to be developed to hold these markets. But in Carpenter's era foreign business continued to revolve primarily around the I.C.I. arrangements, and under Greenewalt Du Pont focused on the domestic nylon market. Equally important, the Executive Committee until 1958 included a solid bloc that consistently opposed any substantial foreign involvement. The dominant figure in this group was Angus Echols, who chaired the Finance Committee after 1940 and had been on the Executive Committee since 1930. Echols had a vivid recollection of Pickard's export debacle in the 1920s and rejected all proposals for foreign investment or trade expansion on the ground that it had "been tried before and didn't work." Another member of the bloc was widely reputed to be suspicious of foreign business because he had been robbed once on a trip to Mexico.[19]

Those who opposed foreign direct investment before 1957, however, had valid reasons for their skepticism. In the immediate aftermath of World War II industrial markets in Europe and Latin America were small, hemmed in by national trade barriers and exchange controls, and prospective return on investments in local enterprises was lower than in the undamaged and growing U.S. economy. Furthermore, direct investments, particularly in Latin America, posed substantial risks. In Brazil, for example, frequent devaluations of the currency diluted Du Pont's earnings from Duperial, and the Executive Committee blocked postwar expansion into heavy chemicals there, fearing that the Brazilian government might adopt a more "realistic" policy and impose major new taxes on foreign owned enterprises. In Argentina the situation was even worse. In 1947 the dictator Juan Peron blocked the expatriation of profits as part of his national development program, and Du Pont feared the eventual confiscation of Ducilo, which had a substantial business in sulfuric acid and rayon, although the Foreign Relations Department accurately predicted that "Peron's plans . . . depend on securing assistance from abroad and there is a limit beyond which he will not go."[20]

At one point in the early 1950s Du Pont contemplated selling off Ducilo to W. R. Grace, which had begun moving into chemicals in Latin America after the war. Greenewalt and Du Pont family members who had been involved in setting up the Duperial ventures were opposed, however, and Grace's offer was considered too small. After Peron's fall in 1954 Ducilo grew rapidly, with an estimated value of over $100 million by the 1970s, which in retrospect justified the view of the Foreign Relations Department that "if you can stay in a market over a period of years, no matter how bad it gets, somehow or other things change around." But the problem of Du Pont investment in regions such as Latin America was made succinctly and pointedly by one of Echols's allies: "In those countries, we are foreigners who come in, make money at their expense and take our profits out. We are always facing oppressive legislation—maybe confiscation."[21]

Circumstances began to change in the mid-1950s. While chemical markets and costs of production at home became less appealing, the prospects for foreign trade and investment brightened. The Eisenhower administration sought to encourage private investment abroad on the assumption that the establishment of financial stability and industrial development, particularly in Europe and Latin America, would utlimately reduce the need for American economic and military aid and thus reduce government spending. To this end tax incentives were offered to U.S. companies investing overseas, and efforts by foreign countries to reduce tariffs and establish full currency convertibility were supported and applauded. A key development in this period was the establishment of the European Common Market, beginning in 1957 with the Treaty of Rome. Meanwhile, in Latin America and other less developed regions, the U.S. tried to fashion aid programs that would create financial stability and discourage the spread of economic nationalism of the Peronist variety.[22]

While the anti-foreign investment bloc on the Executive Committee continued to stymie any major initiatives abroad, Greenewalt became more interested, foreseeing a decline in earnings from U.S. nylon sales. In 1953 Du Pont arranged to license nylon to a Japanese firm, and in the next ten years entered joint ventures there in neoprene and polyethylene plastics. Greenewalt also responded to the continued lobbying of the Foreign Relations Department by authorizing an investigation of investment opportunities in Europe; but the department lacked the staff and technical resources to produce a convincing proposal.

Ironically, the initiative for a major move abroad came from one of the manufacturing departments, and was an indirect result of domestic market problems following the split with I.C.I. When the British company acquired a dye plant in the United States in 1956, John F. Daley, head of the Organic Chemicals Department, decided to retaliate and dispatched a mission to survey the possibilities of establishing a freon venture in the United Kingdom. Daley's group decided that the market for freon was too small, but recommended an investment in a neoprene plant. Synthetic rubber was protected by a high tariff, but manufacturers in Britain could sell to the continental market with few restraints. Furthermore, the British government was anxious to promote industrial development in the depressed area around Londonderry in Northern Ireland, a city located advantageously close to cheap power sources and water and air transportation. In 1958 Du Pont began construction of a neoprene plant near Londonderry that would eventually encompass an orlon mill and various plastics operations for sales in the Common Market.

Once this first step had been taken, the momentum for foreign expansion took hold at Du Pont, fortuitously aided by the retirement of Echols and other Executive Committee members who had blocked these measures in the past. In 1959 Du Pont set up a plastics plant in the Netherlands, expanded later to include orlon and teflon production. In France Du Pont entered a joint venture with Kuhlmann, and in Germany Du Pont took over Adox Fotowerke, an internationally established firm that had pioneered in the development of X-ray film. Du Pont's expansion embraced its traditional area of operation in Latin America as well, with Ducilo moving into nylon and petrochemicals in 1961, and other branches producing auto paints, finishes, and refrigerants in Cuba, Venezuela, Brazil, and Mexico. By 1963 Du Pont's foreign sales, including exports and production abroad, was running at more than $400 million, almost 18 percent of total company sales volume. Total foreign direct investment had doubled in the five years following the establishment of the Londonderry plant, and this expansion was financed largely from earnings from foreign sales.[23]

To manage this sprawling empire, Greenewalt ordered the reorganization of the Foreign Relations Department, now to be called the International Department and given a status equal to the manufacturing departments. Appropriately, Walter Carpenter's son, W. Sam Carpenter III, was appointed head of the reorganized operation. Sam Carpenter not only shared his father's interest in foreign operations but had also worked in the Organic Chemicals Department, the most outward-looking of the manufacturing units. The International Department was organized initially into two geographic divisions, for Latin America and Europe, and two functional staff divisions, one responsible for coordinating foreign trade, the other for developing new ventures abroad and supervising the company's foreign patent arrangements, the traditional role of the old Foreign Relations Department.

A key element in the coordination of the European market, the most profitable region of foreign operations in the 1960s, was the Geneva sales office established in 1959. All Du Pont subsidiaries in Europe sold at cost to Geneva which acted as their distributor, absorbing the taxes on earnings at the relatively modest Swiss levels. The Geneva office was also initially authorized to handle the distribution and sales of Du Pont's U.S. exports to Europe, but this function was to be a source of controversy that reflected deeper divisions within the corporation over the organization of its foreign business.

Before the 1960s the manufacturing departments tended to regard foreign markets as ancillary: when U.S. markets were saturated, foreign sales would be pushed, but as soon as conditions at home improved shipments abroad were cut back, regardless of demand. Similarly, the manufacturing departments were reluctant to support branch manufacturing abroad: the Foreign Relations

Department was never provided with adequate technical assistance in planning ventures in Latin America, and the best managers and technical people were reserved for domestic operations. The International Department after 1958 faced similar difficulties. Although "fully responsible to the executive committee for management and financial return on Du Pont's foreign enterprises," it had "no technical resources of its own and must rely on specific industrial departments for know how and technically trained personnel to run the plants abroad." The departments with the largest foreign commitments insisted on controlling their own export operations and bickered with the International Department over control of foreign branch production.[24]

Relations were complicated further by the reemergence of the tariff issue in the 1960s. Two developments were of particular significance. First, while the formation of the Common Market encouraged direct investment in Europe, it created problems for American exporters as the European countries moved to establish a uniform tariff system that increased overall average rates. Second, the Kennedy administration proposed a major overhaul of international tariff rates in the "Kennedy Round" of the General Agreement on Tariffs and Trade set up after World War II. One major element of the American proposal was a general 50 percent cut in chemical import duties. For their part the Europeans demanded the abandonment of the American Selling Price in organic chemicals that Du Pont had helped fashion in 1922.

A number of Du Pont's manufacturing departments, including the textile and cellophane producers, objected to the Kennedy Round proposals, but Organic Chemicals was particularly irked over the possible end of the American Selling Price. The company decided in 1963 to pull out of the indigo dyestuff field because of foreign competition, but Du Pont wanted to maintain protection of benzene-based dyes, which were held to provide " 'seed' operations for a wide variety of technical developments in organic chemicals." In the end the American Selling Price was preserved after vigorous congressional lobbying by the Synthetic Organic Chemical Manufacturers Association, but substantial tariff reductions were imposed. Corresponding cuts were made by the Europeans, but American chemical firms remained unhappy, since foreign manufacturers paid lower wages and could undercut the Americans in world markets.[25]

The official Du Pont position reflected the interest of the domestic manufacturers. The International Department did not contest this position, since it was clear that measures benefiting European chemical producers would also benefit the Du Pont branch plants, which possessed the same cost advantages over the American-based operations. By the middle of the decade the domestic manufacturing departments were beginning to recognize that overseas operations were to be a major area of growth for the company, a view confirmed by Greenewalt's successor as president, Lammot du Pont Copeland, who noted in 1963: "The course we chart on world markets will determine, in part, the profitability of our enterprises. . . . If we don't participate in the rapidly expanding markets abroad, we will soon be victims of economic isolation."[26]

By 1966 the manufacturing departments were committed to foreign expansion and insisted that they should exercise full control over sales and production, significantly reducing the coordinating function of the International Department. This approach made sense in terms of Du Pont's tradition of decentralization, but it could create problems for production managers with technical experience but little knowledge of foreign legal and political conditions. Sam Carpenter's vow to place Europeans and Latin Americans in top positions in the International Department was undermined when the manufacturing departments began putting their own men in charge of branch plants. An attempt to fire an employee in the Netherlands almost created a political incident. Resentment of Du Pont in foreign lands could take violent forms. The manager of the Londonderry plant was murdered during the Irish troubles of the late 1960s, and the American head of Du Pont's Argentine operations went into exile in neighboring Uruguay in the period of terrorist activity following the return of Peron.[27]

These incidents did not result directly from specific failings on Du Pont's part, but rather reflected the growing sentiment abroad against Americans and multinational corporations. At the same time, the image problems of companies such as Du Pont, and perhaps to a lesser extent their operating problems in foreign marketing, were affected by the shift away from a coordinated international organization. Many diversified industrial firms expanding abroad in this era had to make similar choices, between a system that promoted coordination and control of foreign manufacturing and one that emphasized sensitivity to international markets and politics. Neither strategy has been entirely satisfactory, and companies such as Du Pont in the 1960s, whose foreign commitments have become a substantial and permanent feature of their total operations, have experimented with organizations that combine control of production by domestic departments and coordination of sales and financing through an international department.[28]

While the internal controversy over organization smoldered on, Du Pont's foreign sales and investment continued to grow. Between 1959 and 1972 direct investment abroad increased from $300 million to more than $1.6 billion and foreign sales rose by $640 million, constituting more than one fifth of Du Pont's total sales volume. While overall return on investment sagged during this period and continued to decline in the early 1970s as inflationary pressures and two recessions took their toll, the benefits of foreign operations became increasingly apparent. By 1978 foreign sales comprised 31 percent of Du Pont's $12 billion volume, and return on investment abroad was acknowledged to have been the major reason for the company's recovery from the 1974–75 recession.[29]

Du Pont was not the only U.S. chemical firm to join the scramble for overseas investment. Dow and Monsanto had begun expanding abroad five years ahead of Du Pont, and U.S. chemical investment overseas increased sixfold in the 1960s. By the middle of the following decade all of the major chemical companies except Allied were heavily committed to foreign sales and manufacturing. Dow led the pack with more than 45 percent of sales in foreign markets, followed by Hercules with 40 percent and American Cyanamid and Union Carbide with 33 to 35 percent respectively. In 1963 U.S. chemical sales abroad exceeded $617 billion, about two thirds accounted for by subsidiaries located outside the country. Chemical manufacturers of course were not alone in this expansion: auto makers and producers of electrical equipment and fabricated metals registered an even larger volume of sales and manufacture overseas. Foreign producers, particularly in Europe, were becoming increasingly alarmed over the "American challenge," while the U.S. government began to view with concern the flow of domestic earnings into foreign investments.

Less apparent at the time was the beginning of a "foreign counterattack," particularly from Germany. Britain's I.C.I. continued to be one of the most active chemical firms in international markets, with 30 percent of sales in 1953–62 made by foreign subsidiaries, plus 17 percent in exports. Britain's exclusion from the Common Market in this period encouraged the British to begin substantial direct investment on the Continent; and, as noted earlier, I.C.I.'s move into the United States after the split with Du Pont had been sufficiently threatening to persuade the Wilmington firm to make its own plunge into foreign manufacturing.

The German firms at this point were perceived as less of a threat. Following the breakup of I. G. Farben, the "big Three" of German chemicals—B.A.S.F., Bayer, and Hoechst—had concentrated on rebuilding their plants and technical capabilities, channeling 4 percent of gross earnings into research in the 1950s. But the Germans, as always, had their eyes on foreign markets. Bayer and Hoechst moved quickly back into the Latin American dye and drug markets, reconstructing the prewar sales network and repurchasing plant they had lost in the war, while B.A.S.F. concentrated on heavy chemicals' development. By 1962, with German chemical exports running at $1.2 billion, the companies had reclaimed much of the foreign territory lost in the tumultuous years between 1942 and 1952, and were posed to reenter the U.S. market where the demand for German specialty dyes and pharmaceuticals remained strong. German direct investment in the U.S. more than doubled between 1959 and 1962. By 1979 the $55 million in American assets of Hoechst alone was more than double the total German investment there twenty years earlier, and the $5.4 billion in earnings from sales by American subsidiaries of the "big three" were equal to more than 40 percent of Du Pont's total sales.

The Swiss chemical trio—Ciba, Geigy, and Sandoz—participated in this expansion in the U.S. market, although their growth was not as dramatic. The Swiss firms had set up U.S. branch plants in the 1920s in response to the protective tariff, and had remained relatively untouched by the wartime upheavals accompanying the seizure of the German chemical subsidiaries and the dismantling of I. G. Farben, although in 1950 they dissolved their cartel, the "Basel I.G," partly because of pressure from American antitrust authorities on their American subsidiaries. In 1959 the Swiss companies, with $250 million in U.S. assets, were the largest foreign-owned chemical manufacturers in the United States, with Geigy in particular growing rapidly on the basis of its DDT sales. Throughout the 1960s Ciba and Geigy continued to grow, culminating in a controversial merger in 1970 that established a $1.4 billion company with global sales operations and manufacturing subsidiaries in Italy, Brazil, Britain, and North America. Shrewd, cautious, and persistent, the Swiss were never large enough to challenge the chemical giants, but retained a significant market share in a limited range of products where their technological position was unassailable.[30]

By the 1960s a new international order had begun to take shape in the chemical industry, along lines markedly different from the system that had prevailed up to World War II. In place of the global network of technical agreements and joint ventures, and the intricacies of business diplomacy, the new structure was characterized by parallel networks of branch plants and sales organizations. The companies that had dominated world chemical markets before 1939—Du Pont, I.C.I., the German "big three," and the Swiss—retained their respective positions, but all of them faced increasing competition in traditional markets and rising production costs at home. These factors impelled them into international direct investment, within each other's formerly sacrosanct territories as well as into hitherto underdeveloped regions in Asia and Latin America. From the 1960s through the early 1970s, with the general growth of the world economy, these converging strategies of expansion proved to be mutually satisfactory, with only minor instances of friction and confrontation. In the more troubled years after 1973, however, the companies entered a new era of commercial and technological rivalry on a scale unmatched since the 1920s. For Du Pont, and for the American chemical industry, the challenges of the present bear a peculiar resemblance to those they have faced in the past.

CHAPTER FOURTEEN

Performance and Prospect: Du Pont in the Twentieth Century

In the aftermath of World War I, Du Pont projected an image of dynamism as a company on the leading edge of new technology and a major innovator in the art of managing a complex industrial organization. Following World War II, Du Pont dominated the international chemical scene, the virtual embodiment of American entrepreneurial and managerial capabilities. More recently the company, despite its massive size and continuing significance in a wide variety of industries, has lost much of its glamour. After more than a decade of diminishing earnings per share and huge new debts Du Pont was characterized condescendingly as an "aging blue chip" by Wall Street analysts, whose reaction to the Conoco merger in 1981 was largely negative, and reinforced in the short run by a softening of oil prices. On the other end of the spectrum, critics of big business regarded Du Pont as a "corporate dinosaur," whose size impeded efficiency and whose growth was attributable not to flexibility and innovation, but to abuses of its power over markets and predatory incursions against smaller more innovative competitors.

Du Pont's reputation in the twentieth century parallels and reflects to a certain extent that of the United States. Fueled by innovative industries in the first three decades of the century, the United States survived the depression and emerged from World War II as a dominant element in international markets. But in the past two decades that position has eroded, squeezed by the rising costs of wages, materials, and energy, and challenged from abroad by the revived economies of Western Europe and East Asia.

Clearly Du Pont and the American economy in which it continues to occupy an important position face serious problems in the immediate future and over the long term, for without a major revival of capital investment and technological innovation, losses that seemed temporary and recoverable may signal a permanent deterioration of the competitive position of American industries. At the same time, Du Pont in its own field and American industries in general have substantial resources to draw upon in confronting these challenges. With a skilled and educated work force, a well-developed physical base for research, and an international network of manufacturing and marketing operations, Du Pont remains a formidable enterprise. Whether Du Pont will remain so into the twenty-first century will depend, as in the past, on the resiliency and foresight of its leadership, and, perhaps no less important, on the response of the American economy as a whole to the challenges of the future.

Du Pont's Troubled Empire, 1963–78

While Du Pont's sales continued to climb in the mid-1960s, passing the $3 billion mark in 1966, the company never fully recovered from the loss of its General Motors earnings, due to fluctuations in national and international economic conditions, some bad management decisions, and continued pressure on costs and prices in the fibers, plastics, and other petrochemical fields. Only once between 1964 and 1978 did earnings per share exceed the $10 level reached in 1964, the last year that G.M. earnings were included; and when adjusted for inflation, the figures become even more modest. Return on investment, based now entirely on operating earnings, declined steadily in this fifteen year period, bottoming out at 2.5 percent in the 1975 recession. Return on equity averaged about 13 percent, a reasonable performance given the troubled economic circumstances, but well below the levels to which Du Pont has aspired in the past. The market value of Du Pont shares tumbled as a result, from a high of $261 in 1965 to a low of $84.50 in 1974, although the decline represented in part an adjustment to more realistic values from the artificially high levels reached in the period of the G.M. divestiture. By 1976 Du Pont stock had recovered more than 50 percent of its mid-1960s value, and in 1979 the company arranged a three for one split of common shares. Nevertheless, the days when Du Pont was the fair-haired boy of Wall Street lay well in the past.[1]

A variety of problems plagued Du Pont and the chemical industry during this era. Increasingly frequent shifts in economic conditions troubled the business community in general. After a spurt of growth stimulated by the 1964 tax cut, the economy suffered a series of setbacks of rising magnitude in 1967, 1970, and 1974. To a certain extent Du Pont's increasing overseas sales and manufacturing operations offset difficulties in the home market, but the 1974 recession following O.P.E.C.'s hiking of oil prices affected virtually all the industrial nations. Even before the dramatic rise in energy costs, Du Pont and other American manufacturers had to cope with inflationary pressures on wages and costs of materials. Up to the mid-1970s Du Pont's traditional practice of financing new capital development from retained earnings had helped the company circumvent the additional problem of servicing external debts. As early as 1967, however, Charles McCoy, Lammot Copeland's successor as chief executive of Du Pont, called attention to the fact that rising costs and declining earnings were placing the company in a position where it would soon have to substantially tap the money markets in order to maintain and expand its plants and research facilities. McCoy's successor, Irving Shapiro, initiated in 1973 a deliberate policy of increasing long-term debt to expand plant capacity in anticipation of rising demand for Du Pont's products after the recession.[2]

This move did not escape criticism by observers in the American financial community. Although the company's debt position was by no means unreasonable, its expansion depended to some extent on major growth in the demand for synthetic fibers, since Du Pont continued to rely on this market for more than 30 percent of sales in the mid-1970s. Contrary to expectations, however, fiber markets remained slow even after the 1974 recession as competition increased throughout the world.

Long before Shapiro's tenure, Du Pont's once unquestioned strategists had encountered serious difficulties managing their huge domain. Lammot Copeland had committed the company to massive capital expansion a decade earlier, financed substantially from earnings, leaving Du Pont with a short-term cash problem in the 1967 slump, although continued sales and production in Europe sustained the company through this time of troubles. Even Du Pont's vaunted research organization was guilty of serious miscalculations. In 1964 Du Pont's Fabrics and Finishes Department introduced a new product called "Corfam" that was touted as a moisture proof synthetic substitute for leather in shoes. A portion of the Old Hickory plant was assigned to produce Corfam for forty shoe manufacturers. Both cost and market projections proved faulty, however, and the product encountered consumer resistance as Corfam shoes were not competitively priced and earned a reputation as uncomfortable and inflexible. By 1966 Corfam was losing its potential market to other synthetic leather products (including a rival Du Pont product called "Pattina") and to cheaper vinyl imports from Europe and Japan.

A year later Copeland resigned under fire for a variety of reasons, including an ill-timed reaction to black unrest in Wilmington in the bitter summer of

1967. Although sales picked up in 1968–70, Du Pont decided to pull out of the Corfam field altogether after losing up to $80 million on the venture. Ironically, Corfam technology and machinery was sold to Poland, which was then beginning its own ill-fated move into consumer goods manufacturing.[3]

Under McCoy, Du Pont recovered ground, partly through sales outside the United States, which more than doubled between 1968 and 1973, and partly through a severe retrenchment effort that included cutbacks in some of the consumer goods fields Copeland had initiated and reduction or elimination of obsolete processes and marginal lines, including polyethylene film, black powder and dynamite, and rayon, as well as Corfam. But the company still encountered skepticism in the financial community about its long-term earnings potential, and also had to deal with a new set of problems that confronted the U.S. chemical industry as a whole.

Foreign competition in American chemical and fiber markets had primarily taken the form of imports in the 1960s, and the habitually protectionist Americans had renewed demands for quotas and higher duties during the 1967 and 1970 recessions. During the next decade, however, a new element was added to the situation by the reappearance of substantial foreign, particularly German, direct investment in chemicals in the United States. Up to the late 1960s the German "big three" had focused on exploiting the European market and on exports, but a number of factors induced them to expand branch operations in the American market.

The problem of excess capacity that plagued American producers spread to Europe after 1970. Average sales growth in the Common Market fell by 50 percent from levels reached in the preceding decade. The Germans and Swiss now faced competition from the newly consolidated French and Italian chemical companies, Rhone-Poulenc and Montedison, that were determined to expand production despite falling demand. The growing cries for protection by U.S. manufacturers threatened their exports as well. Furthermore, labor costs in Europe were increasing. Trade unions had acquired a stronger position, demanding legalized limits on layoffs and "co-determination" with management in running big enterprises, as well as higher wages and increased benefits. By contrast, nonunionized workers in the Southern region of the United States provided cost advantages that American manufacturers, including Du Pont, were already exploiting. Finally, the rising value of the German mark against the dollar, and increasingly unpredictable fluctuations in currency exchange rates after 1971, affected the rate of earnings from exports.

Bayer had moved back into branch manufacturing in the United States as early as 1954 when it set up Mobay, a joint venture in polymethane with Monsanto. The American firm provided financing that the undercapitalized but still technically advanced German company could not procure on its own. In 1967 the U.S. Justice Department forced Bayer to buy out Monsanto's interest in Mobay. In 1974 the German company acquired Cutter Laboratories in California, a financially troubled enterprise that specialized in advanced hospital equipment and pharmaceuticals. By that date Bayer had $250 million in assets in the United States, and more than doubled its investment there over the next four years, buying up Allied Chemicals' organic pigments division in 1977 and establishing a new Agfa photographic film branch. Emulating Du Pont's strategy in Europe in the 1960s, Bayer consolidated its properties under a holding company, Rhinechem, controlled by another Bayer subsidiary in Curacao, so that Bayer could draw on the U.S. money market while substantially reducing its foreign tax burden by transferring earnings to its offshore holding company in the Caribbean.

B.A.S.F. began to reenter the United States in 1959, also through the joint venture route in a nylon and acrylic enterprise with Dow. Under Bernhard Timm in the 1960s B.A.S.F., which had acquired a reputation for lethargy and unimaginative management, expanded dramatically, integrating forward into completely novel fields such as computer tapes while also acquiring an oil company in order to guarantee supplies for its petrochemical production. B.A.S.F. moved abroad in a big way, establishing 289 subsidiaries in one hundred countries including the United States, where it took over a medium size American firm, Wyandotte Chemical Company, in 1970. By 1977 B.A.S.F. had the largest stake of the "big three" in direct investment in the United States, with almost $700 million in assets that increased to more than $1.7 billion by 1980.

Hoechst moved into the U.S. market more cautiously, although it had been one of the first of the successors of I. G. Farben to establish an American sales branch in 1953. Building from a base in pharmaceuticals where it was the largest producer in the world by the early 1970s, Hoechst diversified into polyethylene plastics and polyester film and fabrics, including Trevira carpet fiber. Hoechst established one of the largest petrochemical plants in the United States at Bayport, Texas, and acquired a polyester mill from Celanese Corporation in South Carolina in 1978. While Du Pont and I.C.I. continued to dominate the world chemical and fiber industries in terms of net earnings at the end of the 1970s, the German "big three" were each well ahead of them in sales volume, with a growing investment in manufacturing and sales that curtailed their own earnings but placed the Germans in a strong position for exploiting new markets.[4]

The German companies were not directly competitive with Du Pont except in specific fields such as carpet fibers and plastics. Bayer and Hoechst continued to emphasize pharmaceuticals and pesticides, fields Du Pont avoided, and the American firm also scaled down its dyestuffs production in the 1970s. Nevertheless, in a variety of growing fields, such as agricultural chemicals and hospital equipment, the potential for direct confrontation was strong. The German firms poured substantial funds into research and development in an era when the Americans were reducing or stabilizing their commitments in this area. Furthermore, as the "American challenge" to the Germans in the Common Market began to fade, the foreign companies became more aggressive in direct investment in the United States, moving beyond their traditional practice of entry via joint ventures and competing directly with American chemical companies for investment funds in the U.S. money market.[5]

In the early 1970s, however, American chemical manufacturers were less concerned about foreign direct investment and more alarmed over the measures of their own government, particularly in the area of environmental regulation.

Beginning around 1962, with the publication of Rachel Carson's *Silent Spring*, a meticulous and dramatic exposé of the ecological effects of chemical pesticides, a popular movement emerged demanding government action to enforce strict pollution control standards on American industries. These demands were augmented by the growth of the "consumer movement" spearheaded by Ralph Nader and his associates that accused manufacturers of producing unsafe and environmentally hazardous materials, failing to adequately protect the health and safety of workers, and blocking government efforts to overcome these problems through abuse of influence in Congress and the state and federal regulatory agencies. This new tide of reform crested in 1976 with the passage of the Toxic Substances Control Act aimed directly at the chemical industry, accompanied by measures to strengthen the powers of existing regulatory bodies such as the Food and Drug Administration and Environmental Protection Agency, and the establishment of new agencies including the Occupational Safety and Health Agency and a proposed Consumer Product Safety Commission.

The American chemical industry reacted angrily, arguing that the cost of new regulations would damage its competitive position in world markets by diverting more than 10 percent of earnings that should be invested in research and new product innovations. Led by Dow's Carl Gerstacker, the manufacturers commissioned a study of the effects of the Toxic Substances Control Act on their industry, and established the American Industrial Health Council to counter O.S.H.A.'s efforts to monitor the handling of carcinogenous materials in the workplace.

Du Pont endorsed these responses but also sought to overcome criticism through more positive steps. Spokesmen for the company maintained, and critics acknowledged, that Du Pont had a relatively good record in the areas of environmental and occupational health protection, dating back to the 1930s when it had set up one of the first pollution control programs in the industry and established the Haskell Institute, which carried out research on the health hazards of chemical processing. Nevertheless, Du Pont did not escape unscathed from the controversies of this era. The question of the environmental effects of leaded gasoline that had emerged in the 1920s when Du Pont first produced tetraethyl lead, was raised again in 1973 when the E.P.A. ordered the oil and chemical manufacturers to phase out the production of lead-based fuels for automobiles. Du Pont and other ethyl producers contested this ruling, but ultimately lost their case on appeal in the courts in 1976, although the E.P.A. compromised on this issue by arranging for a ten-year period of reduction rather than elimination of leaded fuels from the retail market. Du Pont clashed again with the E.P.A. over an order to ban fluorocarbon spray propellants in 1978, but lost that battle. By that time, however, Du Pont, having invested more than $700 million in environmental control equipment and health research, announced with some satisfaction that the government was beginning to show "positive signs of . . . a flexible approach to environmental regulations."[6]

Apprehending the prospect of increased government intervention, not only in the areas of environmental pollution and occupational safety but also in the familiar arena of antitrust, Du Pont's board in 1973 took the unprecedented step of appointing as chief executive a man who not only had no ties, direct or indirect, to the Du Pont family but who also had no technical training or experience in the chemical industry. Irving Shapiro, a Minnesota-born lawyer, had worked for the U.S. Justice department until lured to Du Pont in 1953 where he played an important part in the extensive litigation that accompanied the G.M. divestiture. To accommodate this unusual arrangement, Du Pont established a dual executive system. Shapiro, as chairman of the board and chief executive officer, was to be "Mister Outside," representing the company in its dealings with government and the public as well as setting company strategy. His vice-chairman and president, Edward R. Kane, a research chemist who had set up the first dacron plant in the 1950s, was to be "Mister Inside," responsible for internal administration of operations. Kane was succeeded in 1980 by Edward Jefferson, another research chemist who two years later became chairman when Shapiro retired.

Shapiro's selection was accompanied by another event that signaled the end of an era at Du Pont. Shortly before his elevation, Shapiro had superintended the liquidation of Christiana Securities which, with its 28 percent equity in the company, had symbolized the role of the Du Pont family in running the corporation. This role had in fact been diminishing since the late 1940s when Walter Carpenter, Jr., had noted with concern the "paucity" of "owner management" on Du Pont's board and executive committee. By 1963, when Irenee du Pont died, virtually all of the "old guard" of active managers or Du Pont relations with holdings in Christiana and Delaware Realty had passed from the scene. Lammot Copeland had been the last chief executive directly related to the family; Charles McCoy, like Greenewalt and Carpenter, was indirectly tied to the family by marriage. In 1974 only five members of the twenty four-person board were linked to the family, including Greenewalt, McCoy, Copeland, and Irenee du Pont, Jr., who was also the only one active in management as a vice-president and member of the Executive Committee. The eclipse of Christiana thus represented no significant alteration in the control of the company—the Du Pont family still held about 25 percent of corporate stock—but did mark the final transformation of Du Pont from the closely controlled family firm to a corporation similar to other large American enterprises of the twentieth century, with management largely divorced from ownership.[7]

Shapiro's term as chief executive of Du Pont began at the lowest point in the company's fortunes and reputation since the depression. As "Mister Outside," he performed effectively, guiding the company through a new round of Federal Trade Commission hearings on the titanium pigments issue, and emerging as a leading spokesman for big business interests in Washington as chairman of the Business Roundtable, an organization of chief executives of large corporations established in 1972 to contest demands for increased government regulation. In 1978 the Roundtable led a successful lobbying effort to torpedo the Consumer Products Safety Commission and helped block the Humphrey-Hawkins bill, a proposal to expand the role of the federal government in job creation programs, the first major initiative in this direction since the New Deal. Business leaders attributed their success to better coordination of lobbying actions and a more flexible stance, emphasizing positive alternatives rather than simple opposition to undesirable legislation and regulation. Critics were inclined to give more weight to such factors as increased corporate contributions to political campaigns and to growing conservatism and disillusionment of the American public with conventional government measures to reduce inflation and restore economic growth. Whatever the case, by the end of the 1970s big businesses were wielding more influence in Washington than at any time since the Eisenhower era.

Meanwhile, Shapiro and Kane began to lay the groundwork for the revitalization of Du Pont. One major objective was to reduce the company's dependence on synthetic fibers and to reorient it toward new fields, reminiscent of Du Pont's diversification strategy at the end of World War I. This process proved more demanding in the 1970s, given the rapid growth of new as well as established firms in promising fields. Nevertheless, some substantial initiatives were taken into specialized products that offered a high rate of return on relatively small levels of capital expansion, including medical precision instruments such as blood analyzers and X-ray film, and agricultural chemicals and specialty plastics for use in such exotic items as bulletproof vests. By 1979, more than one third of Du Pont's net earnings came from sales of these kinds of products.

A second goal for Du Pont's leaders in the 1970s was the reestablishment of some degree of central control over what had come to be regarded as departmental "fiefdoms," again with the intention of shifting away from reliance on traditional product lines. A corporate planning unit was set up to provide the Executive Committee with an overall view of developments in Du Pont's numerous markets and to identify areas for expansion. The research organizations in the company were also subjected to a stringent review to ensure that their operations were oriented "in a commercial direction," with assurances that corporate spending on research would be increased to match the levels of the German firms. To that end, Edward Jefferson, Shapiro's heir apparent, was designated to coordinate Du Pont's research efforts, providing the company's scientists with greater access to top company management than they had experienced since the era of Lammot du Pont.

Du Pont's third major aim in this period of reconstruction was to meet the challenge of rapidly escalating energy costs that confronted the entire industry, and indeed the economy as a whole, in the late 1970s. In the immediate aftermath of the 1973 oil price increases, Du Pont had declined to pass on the costs to its customers, absorbing almost $1 billion in additional expenditures that contributed to its already troubled earnings position. Shapiro later acknowledged reluctantly that Du Pont's restraint in this situation was probably a mistake from a commercial point of view, and proposed to avoid future problems by expanding for the first time into the oil supply field, a course already adopted by Dow and B.A.S.F. In 1976 Du Pont negotiated a billion-dollar joint venture with Atlantic Richfield. When that deal fell through a year later, Du Pont turned to one of Arco's competitors, Conoco, and arranged for a smaller drilling venture off the Texas coast, with the chemical company guaranteed one third of the anticipated supplies for its petrochemical operations. When oil prices were jolted again by the Iranian revolution in 1979, Du Pont began to consider the possibility of acquiring a large oil firm as a long-term solution to its energy needs. While the Conoco takeover two years later was in certain respects a spontaneous reaction to an unanticipated opportunity, it also represented the culmination of years of cautious consideration and experimentation, a process very much in keeping with Du Pont's tradition of deliberation and orientation toward long-term development.[8]

Du Pont and the Chemical Industry: The Legacy

In 1900 Du Pont was a relatively small firm on the American industrial scene, operating in the specialized, although lucrative, field of explosives. Over the past eighty years Du Pont has experienced five distinct stages of growth and change. Between 1902 and 1910, under the leadership of the "triumvirate" of Pierre, Alfred, and Coleman du Pont, the company emerged as the dominant firm in explosives in the United States, merging a number of rivals into a consolidated, vertically integrated unit with a sophisticated system of financial and administrative controls. During the following two decades, despite the internal upheavals generated by an antitrust suit that sundered the explosives company and a bitter struggle for power at the top, Du Pont embarked on a successful strategy of diversification into chemicals, acquiring a variety of smaller companies and some of the most advanced technology available in this rapidly changing industry. This well-managed diversification effort and the introduction of a flexible decentralized system of production and sales in the 1920s helped Du Pont survive the rigors of the Great Depression. In these years the groundwork was laid for Du Pont's transformation into the world's leading producer of synthetic fibers and related polymer-based products in the era following World War II. As competition and costs in these fields increased in the 1950s, Du Pont launched into foreign trade and investment, shifting from its traditional emphasis on domestic markets and becoming a major multinational enterprise. Finally, with the acquisition of Conoco and the move into new fields such as electronics and biomedical products, Du Pont is again being transformed, its future shape and direction as yet unclear.

Throughout much of this era of dramatic changes, certain characteristics of Du Pont in its own policies and operations and in its relations with outsiders have remained constant. While committed to the exploration and exploitation of new developments in technology and new markets, in other respects Du Pont has adhered to established traditions. This continuity reflects in part the role of the Du Pont family, whose members were active in managing the enterprise up to the present generation. It may also be a byproduct of the longevity of the corporation as an institution whose origins date back to the early years of the American republic.

A key factor in the growth of Du Pont has been the orientation of management, both at the highest executive levels and throughout the operating divisions, toward the objectives of the company and the adoption of strategies and organizational measures to fulfill these objectives. Throughout the nineteenth century and well into the twentieth century, Du Pont was directed by a small and relatively cohesive group of family members and close associates who viewed current operations in the perspective of the long-term survival and development of the company. The introduction of centralized financial and administrative mechanisms of control, largely the work of Pierre du Pont between 1903 and 1907, helped establish a system of coherent planning on an enduring basis at a time when the company, like many other family firms, was encountering increasing problems in managing its sprawling domain. The power struggle of 1915–18 resulted in the reinforcement of this system of control by a small body of like-minded owner-managers. Shortly thereafter, the adoption of a decentralized structure at the operational level enabled top management to reduce its role in routine administration while preserving control over the strategic direction of the company as a whole.

Throughout these years, much of Du Pont's growth was achieved through acquisitions of smaller firms and technology, based on a well-developed and comprehensive strategy of diversification. In contrast to the sometimes frenetic activities of some other American businesses in this era, Du Pont moved carefully, seeking to ensure that each new acquisition could be justified on both technical and financial grounds. Mistakes were made: euphoric expectations were not always fulfilled and opportunities were missed. But on the whole Du Pont's long-term approach to diversification paid off, helping to carry it through the depression. Later, the company embarked on an equally cautious and measured advance into foreign direct investment, a strategy that helped cushion Du Pont, although it could not protect the company altogether from the full impact of the decline of the synthetic fiber market in the 1970s.

Du Pont's orientation toward long-range development was rendered possible in part by the flexible but prudent approach to financing new ventures practiced by Pierre du Pont and his successors, but perhaps more importantly by some fortuitous circumstances that influenced the company's financial position. The horizontal integration of the explosives industry achieved by the three cousins in 1902–4 was carried out at a remarkably low cost. Earnings from the sale of munitions during World War I provided the essential "critical mass" that financed the diversification program, including the farsighted acquisition of General Motors stock, whose earnings over the next forty-five years enabled Du Pont to plough back most of its operating profits into new capital investment. Consequently, up to the 1970s the company was able to maintain the tradition begun a century earlier of financing expansion from internal sources, avoiding debt-servicing charges and a dilution of control of the firm and its strategic direction to outsiders. While adherence to this tradition created serious problems after the G.M. divestiture and the cost-price squeeze on the nylon market reduced earnings, Du Pont's experience with "leverage" over the past decade has indicated some of the difficulties that accompany external financing.

As essential to Du Pont's success as long-term strategic planning and internal financing has been its emphasis on research, or more accurately, on the acquisition and control of new technology. Long before the company diversified into chemicals, the Du Ponts recognized the importance of keeping abreast of innovations and developing knowledge. The original Lammot du Pont, founder of Repauno and an inventor in his own right, saw to it that his sons, who would eventually inherit the company, were given the best technical education available at the Massachusetts Institute of Technology. This orientation toward technical innovation was institutionalized at the time the company was reorganized through the establishment of permanent research facilities and reliance on the Development Department to coordinate plans for diversification.

Since World War I, there have been two major shifts in Du Pont's strategy of technological expansion. In the immediate aftermath of the war, as the company began to move into the unfamiliar terrain of chemicals, it relied primarily on the acquisition of technology, particularly from European sources, while Du Pont's research organization focused on refining and improving processes, and Du Pont protected its new position through patent litigation. By the end of the 1920s, Du Pont was undertaking basic research which led to major breakthroughs in the polymer field in the following decade. Although Du Pont's devotion to research diminished somewhat in the postwar years, it continues to be a leader in this field, spending almost twice as much on research and development as its American competitors, and resurrecting the traditional link between research and overall strategic planning as it began to move into biomedical products and other specialty fields in the late 1970s.

While Du Pont's development has been determined in large measure by the skillful deployment of the company's resources, it has also been influenced by external factors. The policies of the United States government have been crucially important in shaping Du Pont's strategy and structure, going back to its earliest years when the first Irenee du Pont negotiated a powder contract with Thomas Jefferson's administration. Abrupt increases and equally sudden disruptions in government demand for munitions stimulated Du Pont's expansion and contributed to the formation of the Powder Trust in the nineteenth century.

The unprecedented demand for explosives by the Allied governments in World War I provided Du Pont with a massive infusion of capital for investment in diversification, and the tariff of 1922, justified largely on grounds of national security, enabled Du Pont to protect its new position in chemicals. Government largesse during the Second World War and Cold War era, although less significant for Du Pont's general strategy and development, continued a long-standing link between the company and the American armed forces.

Du Pont's relations with other agencies of government, particularly the Justice Department, were less friendly. Ironically, however, measures taken by antitrust authorities that Du Pont regarded as hostile and detrimental to its interests at several points led to the redirection of company strategy in ways that ultimately proved beneficial. The divestiture proceedings following Du Pont's first major antitrust confrontation in 1911–12 enabled the company to jettison some of its less desirable properties while retaining control over the smokeless powder field, and concurrent cutbacks in government munitions orders forced Du Pont to move more rapidly toward diversification into chemicals than might otherwise have occurred. The antitrust case also obliged Du Pont to redefine its relations with Nobel and other foreign firms to emphasize patent exchanges rather than market divisions, and helped shape Du Pont's postwar strategy of acquiring foreign technology.

Du Pont's foreign entanglements were a central focus of antitrust action following World War II and led to the breakup of a structure of Anglo-American cooperation in explosives and chemicals that dated back to the turn of the century. Here again measures that seemed at first to disrupt Du Pont's operations proved in retrospect to be less troublesome. Du Pont was able to extricate itself from an increasingly problematical connection with I.C.I., and set the stage for a rapid move abroad in the late 1950s. Meanwhile, the American campaign to break up international cartels and eliminate the war-making capabilities of Du Pont's German competitors removed a potentially formidable rival temporarily from the scene.

Of all Du Pont's antitrust encounters, the General Motors case and subsequent divestiture was probably the most devastating, as Du Pont lost a source of revenues that had cushioned it against the vagaries of the economy for more than a generation. Yet the loss of G.M., after producing a period of general demoralization, forced Du Pont to again review its operations and objectives, and the promotion of foreign direct investment by the U.S. government in this period provided incentives to Du Pont to develop a strategy based on multinational expansion.

Finally, changes in the structure and direction of the international chemical industry influenced Du Pont, which in turn had an impact on developments abroad. Despite the fact that Du Pont's exports and foreign investment constituted a minor element in its operations up to the 1950s, the company's position in the American market, its attitude toward potential foreign competition, and its increasing technical leadership ensured that Du Pont would be a factor that other large international chemical firms had to take into account in their decisions. Du Pont's strategy of consolidation and diversification in the early twentieth century was not unique. By the 1920s, all of the major European chemical manufacturers were moving in the same direction. But while Du Pont and other American producers tended to concentrate on the home market, these large foreign firms were oriented toward exports and investment on a global scale.

Since the latter part of the nineteenth century, however, the foreign chemical manufacturers had constructed an elaborate system of cartels and similar mechanisms for controlling competition and encouraging cooperation. Solvay in heavy chemicals and Nobel in explosives established a network of related enterprises while the Germans dominated the world's dye industry. On the whole, these industrial empires coexisted by remaining within their specialized fields, entering each other's territories only on a limited basis controlled by mutual agreements. The tremendous advances in chemical technology and potential industrial applications in the era of World War I disrupted this system, however, and the European firms sought to redefine relations in a context in which all were pursing expansion through diversification and potential encroachments upon each other's markets.

Du Pont had been drawn into the pre-1914 cartel system through its connections with Nobel, but the 1911 antitrust decision and the onset of war in Europe forced Du Pont to restrict future links to foreign firms to patent and process exchanges. As it turned out, this shift proved beneficial to Du Pont in its diversification program, and it joined the postwar race for expansion through technological innovation. By the end of the 1920s a new balance of power had been established, with Du Pont allied to I.C.I. through a network of patent exchanges and joint ventures, and a similar, although less cordial, relationship with the German giant, I. G. Farben.

The momentum of technical change, in particular Du Pont's increased capabilities and confidence in its research, began to put strains on this balance of power even before World War II wrecked the system. The fabric of cooperation that I.C.I. and Du Pont sought to preserve was shortly to be shredded by Washington's wartime crusade against international cartels. Consequently, when Du Pont and other American chemical firms began to edge back into international operations in the 1950s, the traditional mechanisms of patent agreements and joint ventures were regarded as both unnecessary and legally shaky. There were some joint undertakings, particularly with the Japanese, whose government policies required local participation, but emphasis was placed on direct investment and an extension of Du Pont's internal organization to encompass foreign markets. I.C.I. had already reorganized its international operations along these lines, and the Germans followed a similar course when they reentered foreign markets in force in the 1970s. By the end of that decade a new structure of the international chemical industry could be discerned, characterized by large integrated firms with internally controlled multinational operations in place of market agreements, patent exchanges, and joint ventures.

Du Pont's influence on the development of American business extends far beyond the boundaries of the explosives and chemical industries. The history of the U.S. automobile industry, for example, cannot be fully comprehended without recognizing the crucial role of Du Pont in the restructuring and direction of General Motors for almost half a century. Beyond this direct line of influence, however, Du Pont's impact on American business organization and practice has been remarkable. The company pioneered in the systematic approach to technological research and development and the coordination of research with long-range strategic planning that is at least theoretically observed by most large diversified enterprises. Du Pont provided a model for the organization of massive multidivisional corporations that has long been a central element of major American businesses. Finally, Du Pont was one of the first American businesses to recognize the fundamental interdependence of the world economy, and to design strategies that would take into account technical, organizational, and commercial developments abroad as well as at home. Today Du Pont, like other American businesses—and the American industrial economy—faces serious challenges. Du Pont's potential for survival and future growth in large measure depends on the combination of tradition and innovation that has been the essential feature of its history.

NOTES AND REFERENCES

PROLOGUE

1. Thomas L. Friedman, "Du Pont Victor in Costly Battle to Buy Conoco," *New York Times*, Aug. 6, 1981, p. 1; Robert Metz, "Seagram Role in Du Pont," ibid., Aug. 7, 1981, pt. 4, p. 6; Lee Smith, "The Making of the Megamerger," *Fortune*, Sept. 7, 1981, pp. 58 62.

2. "Du Pont: Seeking a Future in Biosciences," *Business Week*, Nov. 24, 1980, pp. 86–98; "Du Pont–Conoco: Making the Marriage Work," *Chemical Week*, Sept. 2, 1981, pp. 50–56.

3. "The Antitrusters Strike Out," *Chemical Week*, Nov. 5, 1980, p. 12; Charles E. Mueller, "Monopoly and the Law: The Case of the 'Prudent' Monopolist," *Antitrust Law and Economics Review* 11, no. 4 (1979):73–116.

4. James Phelan and Robert Pozen, *The Company State: The Report on Du Pont in Delaware* (New York: Crossman, 1973).

5. Carl Gerstacker, "Chemistry in the 1980s," *Chemical and Engineering News*, Nov. 26, 1979, p. 45.

CHAPTER ONE

1. Herman Mark et al., *Giant Molecules* (New York: Time Life Books, 1966), p. 16.

2. Jasper Crane, vice president, E.I. du Pont de Nemours & Co., to Du Pont Executive Committee, June 30, 1930 (Jasper Crane Papers, Records of E. I. du Pont de Nemours & Co., Inc., ser. 2, pt. 2, Eleutherian Mills Historical Library, Greenville, Delaware; hereafter cited as Crane MS).

3. Frank Sherwood Taylor, *A History of Industrial Chemistry* (London: Heinemann, 1957), pt. 1.

4. Ibid., p. 181.

5. L. F. Haber, *The Chemical Industry During the Nineteenth Century* (Oxford: Clarendon Press, 1958), p. 18; hereafter cited as Haber, *19th Century*.

6. On Lavoisier, see J. R. Partington, *A History of Chemistry* (London: Macmillan, 1962), 3:363–475. On Lavoisier and Irenee du Pont, see William S. Dutton, *Du Pont: One Hundred and Forty Years* (New York: Scribners, 1942), pp. 10–12.

7. Haber, *19th Century*, pp. 25–27, 39–40.

8. Taylor, pp. 182–84; Haber, *19th Century*, pp. 40–41.

9. Haber, *19th Century*, pp. 3–5; Archibald and Nan Clow, *The Chemical Revolution* (London: Batchworth Press, 1952), pp. 91–95, 130–43.

10. T. C. Barker, R. Dickinson, and D. W. F. Hardie, "The Origins of the Synthetic Alkali Industry in Britain," *Economica*, n.s. 23 (1956):158–71; Clow, pp. 60–63, 75–86.

11. Haber, *19th Century*, pp. 15–16.

12. W. J. Reader, *Imperial Chemical Industries: A History*, vol. 1 (London: Oxford University Press, 1970), pp. 44–45, 469–70.

13. Ibid., pp. 48–53, 95–97; J. M. Cohen, *The Life of Ludwig Mond* (London: Methuen & Co., 1956), pp. 126–64.

14. Reader, pp. 106–7; Haber, *19th Century*, pp. 152–57.

15. Haber, *19th Century*, pp. 109–11; on the French chemical industry, see Paul Hohenberg, *Chemicals in Western Europe, 1850–1914: An Economic Study of Technical Change* (Chicago: Rand-McNally, 1967), pp. 78–82.

16. John J. Beer, *The Emergence of the German Dye Industry* (Urbana: University of Illinois Press, 1959), p. 134; O. Steinert and W. Roggersdorf, *Im Reiche der Chemie—100 Jahre BASF*)Dusseldorf: Econ, 1965), p. 69.

17. Haber, *19th Century*, pp. 81–83; Partington, 4:772–94; R. E. Rose, "Growth of the Dyestuffs Industry: The Application of Science to Art," *Journal of Chemical Education* 3, no. 9 (Sept. 1926):973–1007.

18. H. W. Richardson, "Development of the British Dyestuffs Industry Before 1939," *Scottish Journal of Political Economy* 9 (June 1962):110–14; Ivan Levinstein, "Observations and Suggestions on the Present Position of the British Chemical Industry with Special Reference to Coal Tar Derivatives," *Journal of the Society of the Chemical Industry* 5 (1896):351–56.

19. Hohenberg, pp. 35–36. Fuchsin was also known more commonly as magenta.

20. Haber, *19th Century*, pp. 128–36; Beer, pp. 68–100.

21. On tariffs, patents, and German dyestuffs, see Haber, *19th Century*, pp. 199–203, 217–20; Beer, pp. 103–10. On German sales methods, see U.S. Congress, House Committee on Ways and Means, *Hearings on H. R. 2706: Dyestuffs* (Washington, D.C.: Govt. Print. Off., 1919), pp. 101–5; Williams Haynes, *The American Chemical Industry*, vol. 1 (New York: Van Nostrand, 1954), pp. 311–12.

22. On the industrial investment practices of German banks in the late nineteenth century, see H. Neuberger and H. H. Stokes, "German Banks and German Growth, 1883–1913: An Empirical View," *Journal of Economic History* 34 (1974):710–31; H. Neuberger, "The Industrial Politics of the Kreditbanken, 1880–1914," *Business History Review* 51 (Summer, 1977):190–207.

23. Beer, pp. 117–33; Carl Duisberg, *Meine Lebenserinnerungen* (Leipzig: P. Reclam, 1933), pp. 89–95.

24. Haynes, 1:243–44.

25. L. F. Haber, *The Chemical Industry 1900–1930* (Oxford: Clarendon Press, 1971), pp. 175–76; hereafter cited as Haber, *1900–1930*.

26. Haynes, 1:309–13. See also U.S. Tariff Commission, *Census of Dyes and Coal Tar Chemicals 1918* (Washington, D.C.: Govt. Print. Off., 1919), pp. 58–60.

27. Haynes, 1:184–87; Haber, *19th Century*, pp. 142–44.

28. Reader, 1:98–100, 116–23, 291–93; Haynes, 1:271–73.

CHAPTER TWO

1. See the essay on sources for comments on the literature on Du Pont.

2. Dutton, pp. 80–86.

3. On the organization of the Gunpowder Trade Association, see A. P. Van Gelder and H. Schlatter, *History of the Explosives Industry in America* (New York: Columbia University Press, 1927), pp. 126–41; *U.S. v. E. I. du Pont de Nemours & Co., Inc.* 188 *Federal Reporter* 134–39 (1912).

4. Van Gelder and Schlatter, pp. 431–33, 497–502; Dutton, pp. 126–28. The development of dynamite by Nobel is discussed in chapter 3.

5. On Barksdale and Repauno, see Ernest Dale and Charles Meloy, "Hamilton MacFarland Barksdale and the Du Pont Contributions to Systematic Management," *Business History Review* 36 (Summer, 1962):127–52.

6. Williams Haynes, *Cellulose: The Chemical That Grows* (New York: Doubleday, 1953), pp. 30–31, 45–58.

7. Dutton, pp. 158–63; Van Gelder and Schlatter, pp. 875–82. The figures on Du Pont sales and volume for 1914 are from Alfred D. Chandler, Jr., and Stephen Salsbury, *Pierre S. du Pont and the Making of the Modern Corporation* (New York: Harper & Row, 1971), pp. 610–11.

8. Quoted in Chandler and Salsbury, pp. 23–24.

9. Marquis James, *Alfred I. du Pont, The Family Rebel* (Indianapolis: Bobbs-Merrill, 1941), pp. 144–50; Chandler and Salsbury, pp. 49–51.

10. Chandler and Salsbury, pp. 24–33.

11. Ibid., pp. 51–70.

12. Ibid., pp. 85–93, 104–20; see also Alfred D. Chandler, Jr., *Strategy and Structure* (Cambridge: M.I.T. Press, 1962), pp. 57–60.

13. Chandler and Salsbury, pp. 124–30; Haber, *19th Century*, pp. 192–95.

14. Chandler and Salsbury, pp. 110–16. On the early evolution of the Sherman Act, see Richard Letwin, *Law and Economic Policy in America* (New York: Random House, 1965).

15. Mueller, "Du Pont," pp. 96–101.

16. For the Du Pont point of view toward Waddell, see Dutton, pp. 192–93. Gerald Zilg, *Du Pont: Behind the Nylon Curtain* (Englewood Cliffs: Prentice-Hall, 1974), pp. 109–11, follows Waddell's own characterization of the situation. Du Pont felt vindicated by the outcome of a triple damages suit brought by Waddell simultaneously with the antitrust suit that was thrown out of court in 1914. See *Buckeye Powder Co. v. E. I. du Pont de Nemours Powder Co.*, 196 *Federal Reporter* 514 (1914).

17. Chandler, *Strategy and Structure*, pp. 95–96. See also Van Gelder and Schlatter, pp. 804–21 on the development of government smokeless powder facilities.

18. Chandler and Salsbury, pp. 269–71.

19. *U.S. v. E. I. du Pont de Nemours & Co.*, 188 *Federal Reporter* 153 (1912). According to Mueller, "Du Pont," p. 101, in 1907 when the antitrust suit was introduced, Du Pont controlled 94 percent of the production of black powder, 100 percent of smokeless powder production, but only 65 to 70 percent of the production of soda powder and dynamite in the United States. Dynamite sales represented the main source of company profits up to 1915. Consequently the use of the term "monopoly" by the court was somewhat loose, except in the case of black powder and smokeless powder.

20. Chandler and Salsbury, pp. 278–85.

21. On the negotiations leading to the decree, see Chandler and Salsbury, pp. 285–93; and S. N. Whitney, *Antitrust Policies* (New York: Twentieth Century Fund, 1958), 1:192–94. On Hercules Company's smokeless powder development, see Van Gelder and Schlatter, pp. 892–903.

22. Whitney, p. 193.

23. Quoted in Mueller, "Du Pont," p. 108.

CHAPTER THREE

1. Reader, 1:66.
2. Ibid., pp. 126–28, 156–59; Chandler and Salsbury, pp. 170–71.
3. Ibid., p. 159.
4. Ibid., pp. 159–61, 198–99; Chandler and Salsbury, pp. 61, 113–18, 170–73. See also G. W. Stocking and M. H. Watkins, *Cartels in Action* (New York: Twentieth Century Fund, 1946), p. 438.
5. Chandler and Salsbury, pp. 56, 68–104, 142, 159, 181–87; Haynes, 6:130–31.
6. Ibid., pp. 173–76, 188.
7. Memorandum to Executive Committee, Aug. 29, 1916 (Administrative Files, Du Pont Records, ser. 2, pt. 2, box 2, file 40).
8. "Canadian Industries Ltd. and Predecessor and Associated Companies" (Legal Department Papers, Du Pont Records, ser. 2, pt. 2, box 32). See also Reader, 1:204, 207–11.
9. Chandler and Salsbury, pp. 299–300; Reader, 1:173, 196, 208–11.
10. Ibid., p. 299; Stocking and Watkins, pp. 438–39.

CHAPTER FOUR

1. Daniel J. Kevles, *The Physicists* (New York: Alfred Knopf, 1971), p. 137.
2. Information on the dye industry is in the 1918 *Census of Dyes*; on synthetic ammonia, see Haber, *1900–1930*, p. 200.
3. On British dyestuffs development in the war, see Richardson, pp. 112–17; Reader, 1:265–81; Haber, *1900–1930*, pp. 189–93.
4. Haynes, 1:312–14, 2:11–12; Charles E. Munroe and Aida M. Doyle, "Washington's Relations to the Dye Industry Prior to 1914," *Industrial and Engineering Chemistry* 66 (Apr., 1924):417–18.
5. *Census of Dyes* (1918), pp. 62–65; Haynes, 2:247–50.
6. Haynes, *The American Chemical Industry*, vol. 3 (New York: Van Nostrand, 1945), pp. 240–43; Merlo J. Pusey, *Eugene Meyer* (New York: Alfred Knopf, 1974), pp. 118–21; *New York Times*, Jan. 21, 1917, p. 21; Mar. 14, 1918, p. 18.
7. Lammot du Pont to Williams Haynes, Sept. 23, 1944 (Du Pont Records, Administrative Files, ser. 2, pt. 2, box 23; hereafter cited as Admin. Files). A more cautious, and in the short term more accurate view, was expressed by Walter Carpenter, Jr., in a report on dyestuffs to the Du Pont Executive Committee, Dec. 29, 1915. *U.S. v. E. I. du Pont de Nemours, General Motors Corp. et al.*, Defense exhibits (Du Pont) 85; hereafter cited as *U.S. v. G.M.* Additional information on this debate was offered by Irenee du Pont in testimony at the *U.S.* v. *G.M.* trial, Feb. 18, 1953 (*U.S. v. G.M.* Trial Record, pp. 1960–62).

8. Pierre S. du Pont to J. Amory Haskell, Sept. 8, 1916 (Admin. Files, box 1); *New York Times*, Sept. 29, 1917, p. 3, p. 8.

9. J. A. Haskell to Pierre du Pont, Sept. 29, 1916; ibid., Nov. 16, 1916 (Admin. Files, box 1); "Notes on Du Pont–Levinstein Patented Inventions and Secret Processes Agreement, Nov. 30, 1916" (Admin. Files, box 4); *New York Times*, May 1, 1920, p. 1; Mueller, "Du Pont," pp. 146–47.

10. For a dramatic rendering of Crookes's speech, see "Nitrogen," *Fortune*, Aug., 1932, pp. 45–46. Du Pont's Jasper Crane, "Development of the Synthetic Ammonia Industry in the United States," *Industrial and Engineering Chemistry* 22, no. 7 (1930):795, was more skeptical about the significance of Crookes.

11. "American Cyanamid," *Fortune*, Sept., 1940, pp. 102–4; Haynes, 2:80–84. See also Haber, *1900–1930*, pp. 86–90, on the various methods of nitrogen fixation developed before Haber.

12. On Haber's work on ammonia synthesis, see Morris H. Goran, *The Story of Fritz Haber* (Norman: University of Oklahoma Press, 1967), pp. 42–52; and the detailed account in Haber, *1900–1930*, pp. 90–95—the author was the son of Fritz Haber.

13. Haber, *1900–1930*, pp. 198–204, takes issue with the claim made later by the Allies (and repeated by Haynes, 2:57) that Germany stockpiled Chilean nitrates in preparation for war in 1913–14.

14. Chandler and Salsbury, pp. 181–87, 230–37; Haynes, 2:59–60.

15. Irenee du Pont to Pierre du Pont, May 20, 1916; J. B. D. Edge, vice-president, Purchasing Dept., Du Pont Nitrate Co., to C. H. MacDowell, Raw Materials Div., U.S. Council of National Defense, Mar. 16, 1918; Du Pont Nitrate Co., Balance Sheet, Dec. 31, 1918 (Admin. Files, box 3). See also Chandler and Salsbury, pp. 378–79, 400.

16. Irenee du Pont to T. Coleman du Pont, Mar. 17, 1920 (Admin. Files, box 3). See also Chandler and Salsbury, p. 378; Haynes, 2:83–84.

17. Pierre du Pont to Senator Oscar Underwood, Apr. 1, 1916; Senator Willard Saulsbury to Pierre du Pont, Apr. 10, 1916 (Admin. Files, box 3). Saulsbury was related by marriage to the Du Pont family, but had taken sides with Alfred in the dispute over control of the company and had also run as a Democrat against Henry du Pont in 1908 before being elected in 1912.

18. L. E. Edgar, Memorandum on William Beckers, May 18, 1916 (Admin. Files, box 3).

19. Arthur A. Noyes to Pierre du Pont, June 17, 1916; Irenee du Pont to Pierre du Pont, May 20, 1916 (Admin. Files, box 3). See also Haynes, 2:93–97.

20. The nitrogen program was subjected to several Congressional postmortems by the Republicans, anxious to discredit the Wilson administration for waste and incompetence. See U.S. Congress, House, "Muscle Shoals Inquiry," H.D. 119, 66th Cong., 1st sess., 1925; H.R. 998, 66th Cong., 2d sess., 1921; U.S. Congress, Senate, Committee on Agriculture, "Production of Atmospheric Nitrogen," 66th Cong., and sess., 1920; and Haynes, 2:97–103.

21. See Preston J. Hubbard, *Origins of the T.V.A.: The Muscle Shoals Controversy, 1920–1932* (New York: Norton 1968), pp. 4–27, on the Glasgow proposal and the postwar debate over nitrate plants.

22. Crane, "Development of the Synthetic Ammonia Industry," p. 796; Haynes, 4:86–87.

23. Pusey, p. 121.

24. "Union Carbide," *Fortune*, June, 1941, pp. 124–26; Haynes, 4:40–41.

CHAPTER FIVE

1. Chandler and Salsbury, pp. 142, 234–35, 242–44; Dutton, pp. 185–86.
2. Chandler, *Strategy and Structure*, pp. 96–98; Mueller, "Du Pont," pp. 110–11.
3. R. R. M. Carpenter, Development Dept., to Executive Committee, Oct., 1913 (*U.S. v. G.M.*, Govt. exhibit 103).
4. On American Viscose, see D. C. Coleman, *Courtaulds: An Economic and Social History*, vol. 2 (London: Oxford University Press, 1969).
5. Executive Committee to Board of Directors: "Purchase of Arlington Company," Sept. 22, 1915 (*U.S. v. G.M.*, Defense exhibit [Du Pont] 53); testimony of Irenee du Pont, Feb. 17, 1953. *U.S. v. G.M.* Trial Transcript, pp. 1928–29.
6. Chandler and Salsbury, pp. 304–10.
7. Ibid., p. 322.
8. After the Panic of 1907, Pierre du Pont arranged to have corporate bonds and preferred stock listed on the New York Exchange to increase company liquidity, and in 1910 common stock was also listed, although transactions were restricted to shares other than those held by the Du Pont company or family members. See Chandler and Salsbury, pp. 253–54.
9. The details of this episode are reviewed in Chandler and Salsbury, pp. 326–35, and Dutton, pp. 203–15.
10. Chandler and Salsbury, p. 337. Alfred du Pont's case is asserted in James, *Alfred I. du Pont*, pp. 277–83.
11. Dutton, p. 217.
12. Chandler and Salsbury, pp. 342–57.
13. Dutton, pp. 225–28, 254–57; Chandler and Salsbury, pp. 367–71. For an alternative view, see Zilg, pp. 165–67.
14. Development Dept. to Executive Committee: "Hopewell Plant—Progress Report," Jan. 19, 1916; "Report on Utilization of Parlin, Carney's Point and Haskell Plants," May 15, 1916; "Proposed Utilization of Excess Plant Capacities," Dec. 30, 1916 (*U.S. v. G.M.*, Defense exhibits [Du Pont] 81, 54, 59).
15. Chandler and Salsbury, p. 382.
16. Development Department to Executive Committee: "Harrison Brothers and Company," Aug. 11, 1916. *U.S. v. G.M.*, Govt. exhibit 111; Mueller, "Du Pont," pp. 125–26. Through the Harrison Brothers purchase, Du Pont also acquired a part interest in Beckton Chemicals and Cawley Clark Co., makers of lithopone, a zinc-based paint pigment. H. Grubb, General Manager, Paints and Pigments Dept., to Irenee du Pont, Nov. 4, 1921 (*U.S. v. G.M.*, Govt. exhibit 78).
17. John K. Jenney, interview, Columbia University Oral History Project, 1974; hereafter cited as Jenney, Columbia interview.
18. The Du Pont interpretation of the Old Hickory episode was summarized by N. P. Wescott of the Development Dept. in a letter to Raymond L. Buell of the Foreign Policy Association in New York, Mar. 1, 1937 (Papers of Walter S. Carpenter, Jr., Du Pont Company Records, ser. 2, pt. 2, file C-19; hereafter cited as Carpenter MS). See also Dutton, pp. 230–45; Chandler and Salsbury, pp. 410–27. The subject was investigated several times by Congress. For a strong statement of the critics' position see H.R. 998, 66th Cong., 2d sess., 1923. The Nye Committee that probed the arms trade in 1934–35 also resurrected the Old Hickory affair in its investigation of Du Pont.
19. Figures are from Chandler and Salsbury, pp. 607–9.

20. The early phase of Du Pont interest in G.M. is covered in Chandler and Salsbury, pp. 433–43. On the early history of G.M., see Ed Cray, *Chrome Colossus: G.M. and Its Times* (New York: McGraw-Hill, 1980); Lawrence R. Gustin, *Billy Durant, Creator of General Motors* (Grand Rapids: Eerdman, 1973).

21. The debate over G.M. investment is in the Minutes of a Special Joint Meeting of the Executive and Finance Committees, and Minutes of a Special Meeting of the Board of Directors, both on Dec. 21, 1917 (*U.S. v. G.M.*, Defense exhibit [Du Pont] 47); see also Chandler and Salsbury, pp. 450–56.

22. "The History of the Du Pont Company's Investment in the General Motors Corporation," *U.S. v. G.M.*, Defense exhibit (Du Pont) 110; Reader, 1:383–85.

23. On the 1920–21 crisis and reorganization of G.M., see Chandler and Salsbury, pp. 475–91; Chandler, *Strategy and Structure*, pp. 155–71.

24. Mueller, "Du Pont," pp. 373–75. See also Theodore P. Kovaleff, "Divorce American Style: the Du Pont–General Motors Case," *Delaware History* 18, no. 1 (1978):29–31.

25. Mueller, "Du Pont," pp. 141–45.

CHAPTER SIX

1. Haber, *1900–1930*, pp. 314–15.

2. On Irenee du Pont, see Jenney, Columbia interview; Zilg, pp. 245–46.

3. "Du Pont Executive and Finance Committee Members, 1915–1948," *U.S. v. G.M.*, Govt. exhibit 118; Dutton, pp. 187, 214–15, 272–73.

4. Chandler, *Strategy and Structure*, pp. 70–72.

5. Grubb to Irenee du Pont, Nov. 4, 1921.

6. Chandler, *Strategy and Structure*, pp. 116–37; Ernest Dale, "Du Pont: Pioneer in Systematic Management," *Administrative Science Quarterly* 2 (June, 1957):35–40.

7. Mueller, "Du Pont," pp. 142, 275.

8. Haynes, 4:38–9; E. T. Penrose, "The Growth of the Firm: A Case Study—The Hercules Powder Company," *Business History Review* 34 (Winter, 1960):6–13.

9. "American Cyanamid," *Fortune*, Sept., 1940, pp. 102–4; "Union Carbide," *Fortune*, June, 1941, pp. 126–27.

10. "Allied Chemical and Dye," *Fortune*, Oct., 1939, pp. 47–49; Haber, *1900–1930*, p. 350.

11. On the diffusion of Du Pont's approach to organization, see Chandler, *Strategy and Structure*, pp. 324–26, 333–34, 350–51.

12. Mueller, "Du Pont," pp. 191–92. A more detailed examination of the foreign patent negotiations is in chapter 9.

13. Haynes, *Cellulose*, pp. 210–15; Dutton, pp. 296–300. See also "Du Pont: An Industrial Empire," *Fortune*, Dec., 1934, p. 168, on Du Pont's patent litigation against Glidden involving Duco in 1931.

14. Testimony of Irenee du Pont, Mar. 2, 1953 (*U.S. v. G.M.* Trial Transcript, pp. 2331–50); Memorandum of conference, Du Pont and G.M., Aug. 11, 1924 (Admin. Files, box 6, file C-15). See also Stuart Leslie, "Thomas Midgley and the Politics of Industrial Research," *Business History Review* 44 (1980):482–86; G. S. Gibb and E. H. Knowlton, *History of Standard Oil (New Jersey): The Resurgent Years, 1911–1927* (New York: Harper, 1956), pp. 539–41.

15. Charles A. Meade to Irenee du Pont, Sept. 17, 1922; ibid., Oct. 31, 1924; Report of the Committee Appointed by the U.S. Surgeon General, "Health Hazards in the Sale and Use of Ethyl Gasoline," Jan. 17, 1926 (Admin. Files, box 6, file C-15). See also Zilg, pp. 213–18; Gibb and Knowlton, pp. 541–42; Haynes, 4:398–403; Silas Bent, "Deep Water Runs Still," *Nation*, July 8, 1925, pp. 62–64.

16. "Du Pont: An Industrial Empire," pp. 164, 167; "Swiss Family Dreyfus," *Fortune*, July, 1933, pp. 51–55; Haynes, *Cellulose*, pp. 138–45.

17. Haynes, 4:38–39.

18. Mueller, "Du Pont," p. 280; Haber, *1900–1930*, pp. 313–14.

19. B. D. Beyea to Fin Sparre, Development Dept., Nov. 23, 1927 (Carpenter MS, box 823, file C-44); Haynes, 4:233–34.

20. Walter Carpenter, Jr., to T. S. Grasselli, July 31, 1928 (Carpenter MS, box 823, file C-44).

21. Carpenter to Finance Committee, June 15, 1927: H. B. Rust, Koppers Co., to J. P. Morgan, Jr., July 25, 1927; A. B. Echols, Treasurer, to Irenee du Pont, Mar. 16, 1928; U.S. Federal Trade Commission, *Report on Du Pont Investments*, Feb. 1, 1929 (Carpenter MS, box 824, U.S. Steel File).

22. Howard and Ralph Wolf, *Rubber* (New York: Covici-Friede, 1936), pp. 462–63; G. W. Stocking and M. W. Watkins, *Cartels in Action: Case Studies in Business Diplomacy* (New York: Twentieth Century Fund, 1946), pp. 68–71.

23. Irenee du Pont to Carpenter, Sept. 14, 1928; Carpenter to Finance Committee, Sept. 15, 1928 (Carpenter MS, box 824, U.S. Rubber File). See also Chandler, *Strategy and Structure*, p. 434–35; Zilg, pp. 230–31.

CHAPTER SEVEN

1. Haber, *1900–1930*, pp. 279–309.

2. Ibid., p. 279.

3. Quoted in Reader, 2:32.

4. See Richard A. Lauderbaugh, *American Steel Makers and the Coming of the Second World War* (Ann Arbor: U.M.I. Research Press, 1980), pp. 171–95; Ervin Hexner, *The International Steel Cartel* (Chapel Hill: University of North Carolina Press, 1946), pp. 90–91, 206–16.

5. Beer, p. 139. See also Haber, *1900–1930*, pp. 280–81, and other works on I. G. Farben discussed in the essay on sources.

6. Thomas P. Hughes, "Technological Momentum in History: Hydrogenation in Germany, 1898–1933," *Past and Present* 44 (Aug., 1969):110–14.

7. Joseph Borkin, *The Crime and Punishment of I.G. Farben* (New York: Macmillan, 1978), pp. 28–34, 40–44. In 1926 I.G. and Kuhlmann negotiated a cartel agreement over dyestuffs. See Haber, *1900–1930*, pp. 303–4.

8. Haber, *1900–1930*, pp. 248–50.

9. See Robert Brady, *The Rationalization Movement in German Industry* (London: Cambridge University Press, 1934).

10. Wolfram Fischer, "Dezentralisation oder Zentralisation—kollegial oder autoritäre Fahrung? Die Auseniandersetzung um die Leitungsstrucktur beider Enstrehung des IG Farben Konzerns," in *Law and the Formation of the Big Enterprise in the Nine-*

teenth and Early Twentieth Centuries, ed. N. Horn and J. Köcka (Gottingen: Vandenhoek & Rupprecht, 1979), pp. 476–86.

11. In German corporations, there are two boards: the Aufsichsrat, which generally represents shareholders; and the Vorstand, which is directly responsible for management. There is no direct legal parallel in the U.S., but the relationship is roughly similar to that of the Board of Directors and the Executive Committee at Du Pont.

12. I. G. Farben's connections with the Nazi regime in the 1930s are discussed in chapter 12.

13. Reader, 1:310–14, 2:239–47.

14. Ibid., 1:379–85. Dollar figures are in current exchange values, which understates the firm's asset value when Britain returned to the gold standard in 1925.

15. Ibid., 1:390–96. The development of Du Pont–Nobel relations is discussed in more detail in chapters 9–10.

16. Ibid., 1:415–19.

17. Ibid., 1:425–49.

18. Alkasso (the United States Alkali Export Association) was an organization formed under the Webb-Pomerene Act of 1918 that allowed companies engaged in exports to join trade associations that coordinated operations without running afoul of the antitrust laws. Alkasso, however, was later prosecuted for having abused the privilege. See *U.S. v. U.S. Alkali Export Association* 86 *Federal Supplement* 59 (1949).

19. Reader, 1:451–66; Haber, *1900–1930*, pp. 291–302. On Brunner-Mond's development of synthetic ammonia, see G. P. Pollitt, "The Synthetic Ammonia Industry," *Transactions of the Second World Power Conference* (Berlin: V.O.I. Verlag, 1930), 2:145–64.

CHAPTER EIGHT

1. Quoted in Stocking and Watkins, p. 448.

2. Quoted in "Du Pont–IG Relationship," Feb., 1943 (Papers of Roy Prewitt, box 13, Records of the Federal Trade Commission, Record Group 122, U.S. National Archives, Washington, D.C.; hereafter cited as Prewitt MS).

3. Howland H. Sargeant, "The Patent Policy of the Alien Property Custodian of the United States After World War I," Aug. 28, 1943 (Records of the Alien Property Custodian, Records of the U.S. Justice Department, Record Group 60, U.S. National Archives); appellant's trial brief, *U.S. v. The Chemical Foundation Inc.*, Records and Briefs of the United States Supreme Court, 1925; Haynes, 3:259–63.

4. Irenee du Pont, "Memorandum Concerning Case—*U.S. v. Chemical Foundation*," July 1922; Irenee du Pont to T. Coleman du Pont, July 22, 1922 (Admin. Files, Chemical Foundation File); *U.S. v. The Chemical Foundation* 294 *Federal Reporter* 300, 307; 272 *U.S.* 1–28 (1925). Subsequent relations between Du Pont and the Chemical Foundation were not altogether smooth. In 1926 the Foundation threatened a lawsuit against Du Pont on the grounds that its Claude ammonia process infringed the Haber-Bosch patents the Foundation had seized during the war. Du Pont, however, defied the Foundation, which lacked the financial resources (including contributions from Du Pont) to pursue the matter. See Lammot du Pont to W. W. Buffum, Chemical Foundation Inc., Jan. 23, 1928 (Admin. Files, Chemical Foundation File).

5. Zilg, pp. 178–79.
6. Dutton, p. 290.
7. W. C. Spruance, "Memorandum: Acquisition of the 'Know-how' in Dyestuffs Business," July 22, 1921 (Admin. Files, file 110-B).
8. Eysten Berg, "Nitrogen Industry of Germany," Oct. 28, 1919 (Admin. Files, box 5, file 115).
9. Charles Meade to Walter Carpenter, Jr., Nov. 28, 1919 (Admin. Files, box 5, file 115).
10. "Minutes of Meeting Held Nov. 20–22, 1919 at Baur au Lac Hotel, Zurich, Switzerland"; Eysten Berg, "Our Negotiations with Badische," Feb. 3, 1920 (Admin. Files, box 5, file 115).
11. Berg, "Memo on Badische Negotiations," June 29, 1920 (Admin. Files, box 5, file 115).
12. Spruance, "Acquisition of 'Know-how' "; Borkin, pp. 39–40.
13. Berg to Fin Sparre, June 15, 1920 (Admin. Files, box 5, file 115); Mueller, "Du Pont," pp. 177–78.
14. *Census of Dyes*, 1918, pp. 66–73.
15. Hearings on H.R. 2706: Dyestuffs, pp. 1–13, 314–24; Charles K. Weston, Du Pont Publicity Bureau to R. R. M. Carpenter, Oct. 23, 1919; Report of Committee on Legislation, American Dye Instutute, April 1, 1921 (Admin. Files, box 4, file 99); *New York Times*, Sept. 27, 1919, p. 12.
16. Irenee du Pont to Senator Philander Knox, Oct. 15, 1919 (Admin. Files, box 4, file 99).
17. C. K. Weston to Charles Meade, Dec. 10, 1920; Thomas R. Shipp, Du Pont Publicity Bureau, "Memorandum: Our Present and Proposed Activities," Sept. 12, 1921; U.S. Congress, Senate Special Committee to Investiage the Munitions Industry, *Munitions Industry Hearings: E. I. du Pont de Nemours & Company* (Washington, D.C.: Govt. Print. Off., 1936), exhibits 909, 912; Frank Taussig, "The Tariff Act of 1922," *Quarterly Journal of Economics* 37 (Nov. 1922):17–18.
18. Meade to Irenee du Pont, July 20, 1921 (Admin. Files, box 5, file 110-B).
19. M. R. Poucher, "Dye Legislation Memorandum," Apr. 17, 1922 (Admin. Files, box 4, file 99); Taussig, pp. 18–19.
20. Jules Backman, *Foreign Competition in Chemicals and Allied Products* (Washington, D.C.: Manufacturing Chemists Association, 1965), pp. 20–25; Haynes, 3:269–72. On protective tariff laws on dyes and chemicals in other countries, see *Census of Dyes*, 1924; and Haber, *1900–1930*, pp. 238–46.
21. U.S. Congress, Senate Judiciary Committee, *Hearings on Alleged Dye Monopoly* (Washington, D.C.: Govt. Print. Off., 1922); Haynes, 3:272–73.
22. Irenee du Pont, Memorandum on Meeting with M. R. Poucher and F. P. Garvan, Nov. 9, 1923; F. W. Pickard to Irenee du Pont, Jan. 1, 1924 (Admin. Files, box 6, file C-15).
23. Reader, 2:410–12, "Du Pont–IG Relationship."
24. Irenee du Pont to F. W. Harrington, Mar. 6, 1926 (Admin. Files, box 6, file C-15); Haynes, 4:233–36.
25. Quoted in Mueller, "Du Pont," p. 178. Du Pont subsequently bought Niagara Ammonia and in 1930 acquired Roessler and Hasslacher, a diversified chemical company with an ammonia plant at Niagara Falls. By 1936, Du Pont had a 36 percent share of the U.S. nitrogen market, second only to Allied Chemical.

26. "Du Pont–IG Relationship."
27. Stocking and Watkins, pp. 405–6, 477–78.

CHAPTER NINE

1. Leonard Yerkes to Walter Carpenter, Jr., Aug. 20, 1919 (Admin. Files, box 5, file 115).
2. Haynes, 4:376–78; Dutton, pp. 302–5.
3. Jenney, Columbia interview; Dutton, pp. 309–11; Walter Carpenter, Jr., to Executive Committee, Jan. 4, 1927; C. H. Biesterfeld, Legal Dept. to C. R. Mudge, May 26, 1927 (Carpenter MS, box 818, Rayon File).
4. Charles Meade to Carpenter, Nov. 29, 1919; Fin Sparre to M. R. Poucher, Dec. 8, 1923 (Admin. Files, box 5, file 115; box 19, Poucher/Meade File); Haynes, 4:87–88; Dutton, pp. 335–36.
5. H. G. Haskell to J. E. Crane, July 10, 1923; Haskell to Lammot du Pont, June 27, 1922 (Papers of the Foreign Relations Department, Du Pont Records, ser. 2, pt. 2, box 537; hereafter cited F.R.D. MS).
6. Sir Harry McGowan to Pierre du Pont, Sept. 7, 1917; Pierre du Pont to McGowan, Oct. 15, 1917; McGowan to Pierre du Pont, Nov. 20, 1917 (Admin. Files, box 2, file 40-E).
7. Stocking and Watkins, p. 439; Reader, 2:38.
8. Reader, 2:212–18; G. D. Taylor, "Management Relations in a Multinational Enterprise: The Case of Canadian Industries Ltd., 1928–1948," *Business History Review* 55 (Autumn, 1981):337–58.
9. R. R. M. Carpenter to Pierre du Pont, Mar. 31, 1917 (Admin. Files, box 2, file 40).
10. R. R. M. Carpenter to Pierre du Pont, Aug. 16, 1918; Pierre du Pont to Irenee du Pont, Mar. 11, 1919; F. W. Pickard to Pierre du Pont, Mar. 15, 1919; R. R. M. Carpenter to Pierre du Pont, Mar. 25, 1919; ibid., Apr. 11, 1919; Pierre du Pont, Circular letter, Apr. 16, 1919; May 19, 1919 (Admin. Files, box 1, file 1; box 2, file 40-D).
11. Jenney, Columbia interview.
12. "Memorandum of Discussion with the British Representatives," Apr. 7, 1919; Lammot du Pont to Pierre du Pont, Jan. 29, 1918; R. R. M. Carpenter, Memorandum of Conversation with Paul S. Reinsch, Aug. 16, 1918; Lammot du Pont to Irenee du Pont, Oct. 22, 1921; F. W. Pickard to Charles Meade, Oct. 25, 1921 (Admin. Files, box 1, file 1; box 2, file 40-D; box 17, China file).
13. Irenee du Pont to Charles Meade, Oct. 6, 1921 (Admin. Files, box 17, China file).
14. Reader, 2:27, 34, 110, 415–17; Wendell Swint to Lammot du Pont, Feb. 28, 1933 (F.R.D. MS, box 533).
15. Stocking and Watkins, p. 450; John K. Jenney to Lammot du Pont, Feb. 23, 1929 (F.R.D. MS, box 527).
16. John K. Jenney to J. E. Crane, "Memorandum on Foreign Activities of the Du Pont Company," Dec. 9, 1936 (Crane MS, box 1039).
17. Jenney, Columbia interview.
18. Reader, 2:40–54.

19. Stocking and Watkins, p. 451.
20. Jenney, Columbia interview.
21. Ibid.; Reader, 2:50–54.
22. Stocking and Watkins, p. 451.
23. Katherine Voorhees, Memorandum on Explosives, Jan. 2, 1942; "Du Pont–IG Relationship" (Prewitt MS); Walter Carpenter, Jr., to Finance Committee, Sept. 2, 1926 (Carpenter MS, box 820, file 13).
24. E. G. Robinson, General Manager, Dyestuffs Dept., to Executive Committee, May 18, 1931 (Carpenter MS, box 822); Stocking and Watkins, p. 452.
25. Jenney, Columbia interview; Reader, 2:216–18.
26. Haynes, 4:378–81, 547; Haber, *1900–1930,* p. 314; Leonard Yerkes to Walter Carpenter, Jr., Nov. 26, 1926; Fin Sparre to Yerkes, July 25, 1927; Pickard to Carpenter, Nov. 26, 1926 (Carpenter MS, box 818, Rayon file).
27. Edmond Gillet to Henry Blum, Aug. 3, 1927; Walter Carpenter, Jr., to Yerkes, June 11, 1926; Yerkes to Carpenter, June 15, 1926 (Carpenter MS, box 818, Rayon file); Jenney, Columbia interview; Haynes, 4:383.
28. See chapter 11 on Du Pont's research in polymerization that led to the development of nylon.
29. Jenney, Columbia interview; Jenney, "Memorandum on Foreign Activities of the Du Pont Company"; Wendell Swint to Lammot du Pont, Feb. 28, 1933 (F.R.D. MS, box 533).

CHAPTER TEN

1. Sir Harry McGowan to Walter Carpenter, Jr., Feb. 2, 1931 (Carpenter MS, box 820, file 5).
2. Wendell Swint to G. H. White, I.C.I. (New York), Mar. 5, 1935; Swint to Executive Committee, Mar. 6, 1935 (F.R.D. MS, box 527).
3. Lammot du Pont to H. J. Mitchell, Jan. 14, 1937; Mitchell to Lammot du Pont, Feb. 5, 1937 (Carpenter MS, box 820, file 6).
4. Wendell Swint, "Chemical Industry in Argentina," Feb. 26, 1931; Swint to Foreign Relations Committee, "Possible Chemical Manufacture in Argentina," June 8, 1931 (Crane MS, box 1039).
5. John K. Jenney, "Memorandum on Foreign Activities of the Du Pont Company."
6. "Merging of Argentine Interests," Feb. 7, 1934; J. E. Crane to Lammot du Pont, July 6, 1933 (Crane MS, box 1039).
7. Crane to Jenney, Dec. 2, 1936; Robert Salmon to Duperial Shareholders Committee, Nov. 12, 20, 1936; I.C.I., "Executive Memorandum on Duperial Argentina," July, 1937; Salmon to Crane, Sept. 7, 1936; Salmon to Adolfo Hirsch, April 24, 1937; Hirsch to Salmon, April 26, 1937; Salmon to Hirsch, April 29, 1937; Salmon, Memorandum on Bunge and Born, July 22, 1937; Swint to Foreign Relations Committee, Apr. 1, 1938; Swint, "I.C.I.'s Proposal to Effect a Merger with Bunge and Born," May 12, 1939 (Crane MS, box 1029).
8. Swint, "Proposed Consolidation of Du Pont and ICI Brazilian Interests," Mar. 23, 1936; W. H. Coates, "The Duperial Companies: Argentina and Brazil," July 5,

1938; E. E. Lincoln to Swint, "Outlook for American Investment in Brazil," Feb. 26, 1942; ibid., Nov. 6, 1942 (Crane MS, box 1036).

9. "Du Pont–IG Relationship."

10. Ibid.

11. Ibid.

12. John K. Jenney to Executive Committee, Mar. 1, 1934 (Admin. Files, box 17, Foreign Relations Dept. file); Stocking and Watkins, pp. 110–12, 469–70, 488–89.

13. See chapter 12 on I.G.'s arrangements with Standard Oil (New Jersey) in the synthetic fuel and synthetic rubber fields.

14. "Du Pont–IG Relationship."

15. Stocking and Watkins, p. 122.

16. J. E. Crane to Lord Melchett, Dec. 12, 1938 (Carpenter MS, box 820, file 7).

CHAPTER ELEVEN

1. John Jenney, Columbia interview.

2. Walter Carpenter, Jr., to F. E. Williamson, Public Relations Dept., May 20, 1941 (Carpenter MS, box 833, file 24).

3. Figures are from "Twenty Five Year Financial and Operating Report," Du Pont, *Annual Report* (1945), pp. 40–41. In addition to a regular bonus plan for employees, Du Pont introduced a stock option plan for top executives in 1927, selling about 240,000 shares or 2 percent of the outstanding stock in the firm on the option basis in 1927–32.

4. David C. Kyvig, *Repealing National Prohibition* (Chicago: University of Chicago Press, 1979), pp. 78–81; Elliot A. Rosen, *Hoover, Roosevelt, and the Brains Trust* (New York: Columbia University Press, 1977), pp. 27–37, 243–68, 386. See also George Wolfskill, *Revolt of the Conservatives: A History of the American Liberty League* (Boston: Houghton Mifflin, 1962).

5. On the Nye committee investigation, see John E. Wiltz, *In Search Of Peace: The Senate Munitions Inquiry* (Baton Rouge: Louisiana State University Press, 1963), pp. 76–81, 122–26; and a rather different version offered by Zilg, pp. 309–14.

6. Du Pont, *Annual Report* (1936), p. 17; ibid., 1937, pp. 20–21.

7. Du Pont "welfare" measures are reviewed in *Annual Reports* for 1934, 1936, and 1937. See also Harold C. Livesay, *American Made: Men Who Shaped the American Economy* (Boston: Little, Brown, 1979), pp. 183–87, for personal reminiscences about labor relations at Du Pont in the 1930s.

8. On Du Pont research, see E. K. Bolton, "Du Pont Research," Address to the Society of the Chemical Industry, Jan. 5, 1945; Lawrence Lessing, "Du Pont: How to Win at Research," *Fortune*, Oct. 1950, pp. 115–18, 122–34. On polymer research generally, see Herman Mark, *Giant Molecules*, pp. 73–81, 99–107; and C. Freeman et al., "The Plastics Industry: A Comparative Study of Research and Innovation," *National Institute Economic Review* (London), no. 26 (Nov., 1963), pp. 22–33.

9. H. T. Rutledge, "Air, Water, Coal = Hosiery," *Scientific American* 162 (Feb., 1940):78–81; "Nylon Brews a Textile Revolution," *Fortune*, July, 1940, pp. 56–60, 114–16; James A. Lee, "Nylon," *Chemical and Metallurgical Engineering*, Mar., 1946, pp. 96–99. Mueller, "Du Pont," pp. 312–14, takes issue with Du Pont's claim

to have spent $6 million on research and development of nylon, computing a total research investment cost of $2 million on the basis of materials presented in the I.C.I. case.

10. Richard Hewlett and Oscar Anderson, *The New World, 1939–1946* (University Park: Pennsylvania State University Press, 1962), pp. 105–12; C. A. Rittenhouse, "Measures to Protect the Company's Interests in Connection with the Construction and Operation of Hanford Engineer Works," Oct. 13, 1943; Walter Carpenter, Jr., "Comments," Oct. 18, 1943 (Carpenter MS, box 830, file 5.1).

11. C. M. Stine to Executive Committee, Feb. 12, 1945 (Carpenter MS, box 830, file 5.1); Du Pont, *Annual Report,* 1950, pp. 17–18.

12. Du Pont's struggles with the tax man can be traced in *Annual Reports* for 1942–44 and 1952–53. On the history of the excess profits tax laws, see Sidney Ratner, *Taxation and Democracy in America* (New York: John Wiley, 1967), pp. 506–8, 525–27.

CHAPTER TWELVE

1. Quoted in Reader, 2:198–99.

2. Taylor, "Management Relations in a Multinational Enterprise," pp. 343, 351.

3. H. J. Mitchell to Lammot du Pont, Feb. 5, 1937; Memorandum, " '66' Polyamide," July 28, 1938; C. S. Robinson, I.C.I. (New York) Ltd., to Fin Sparre, Dec. 9, 1938. *U.S. v. I.C.I.,* Defense exhibits D-1134 (Du Pont), D-1141 (Du Pont), D-1148 (I.C.I. 782); Reader, 2:370–72.

4. P. C. Allen to W. F. Luytens, "Exchange of Information of Polythene with Du Pont's," June 18, 1942; W. S. Carpenter, Jr., to Lord McGowan, Sept. 24, 1942. *U.S. v. I.C.I.*, Govt. exhibit 669 (I.C.I. 918), Govt. exhibit 672; Reader, 2:433–34; J. E. Crane to Carpenter, Sept. 10, 1942 (Crane, MS, box 1043).

5. Reader, 2:415–416; Wendell Swint, "Commercial Relations with I.C.I.," Dec. 20, 1939 (Crane MS, box 1043).

6. On Arnold, see Mark Lytle, "Thurman Arnold and the Wartime Cartels," (MS 1976, Yale University, Sterling Memorial Library); Gene M. Gressley, "Thurman Arnold, Antitrust and the New Deal," *Business History Review* 38 (Summer, 1969):214–31.

7. "Box Score on Antitrust Suits," Apr. 11, 1955 (Legal Dept. MS, box 851).

8. Robert De Right, Foreign Relations Dept., to Wilfred Wallace, Export Manager, Nylon Division, Du Pont, Feb. 5, 1947; C. H. Greenewalt to Executive Committee, Report of London Conference with I.C.I. Ltd., June 4, 1948; *U.S. v. I.C.I.*, Defense exhibit D-1168 (Du Pont), D-933 (Du Pont); Reader, 2:436–38.

9. *Business Week*, Dec. 21, 1956, pp. 98–100; Reader, 2:440–41; Jenney, Columbia interview.

10. See essay on sources on I. G. Farben's development of synthetic fuels and synthetic rubber.

11. Borkin, *I. G. Farben,* pp. 66–75; Arthur Schweitzer, *Big Business and the Third Reich* (Bloomington: University of Indiana Press, 1964), pp. 540–43.

12. H. M. Larson, E. H. Knowlton, and C. S. Popple, *History of Standard Oil (New Jersey): New Horizons 1927–1950* (New York: Harper & Row, 1971), pp. 153–57; Reader, 2:166–70; Du Pont Executive Committee Minutes, June 11, 1930; Frank

Howard, vice-president, Standard Oil (New Jersey), to Jasper Crane, July 10, 1930 (Crane MS, box 1038).

13. Larson et al., pp. 170–72; Borkin, pp. 76–78.

14. Arthur Murphy, Foreign Exchange Department, Treasurer's Office, Du Pont, to H. H. Ewing, Foreign Relations Dept., London, May 25, 1934; Henry E. Ford, Foreign Relations Dept., London to Murphy, July 10, 1934 (Crane MS, box 1038); "Memorandum Re IG Investment" (Carpenter MS, box 833, file 24).

15. Stocking and Watkins, pp. 501–5; Borkin, pp. 80–94; Larson et al., pp. 433–43.

16. Stocking and Watkins, pp. 402–4; W. R. Hutchinson, Antitrust Division to Thurman Arnold, May 12, 1942; Henry Stimson, U.S. Secretary of War, to Attorney General Francis Biddle, Aug. 29, 1942; Biddle to Stimson, Sept. 2, 1942; Wendell Berge, Antitrust Division, to Theodore Metcalf, Nov. 22, 1946 (Records of U.S. Justice Department, *U.S. v. Rohm and Haas et al.*, file no. 60–374–2).

17. Abstract of Proceeding before the Securities and Exchange Commission, "In the Matter of Investment Trust Study, American IG Chemical Corporation," Feb. 4, 1938 (Crane MS, box 1038); Herman Schein, "History of the Vesting of General Aniline by the Office of the Alien Property Custodian," U.S. Office of the Alien Property Custodian, Aug., 1945, U.S. National Archives, Record Group 131; Borkin, pp. 186–90, 194–222.

18. Borkin, pp. 135–56. See also Josiah E. Du Bois, *The Devil's Chemists* (Boston: Beacon Press, 1952), passim.'

19. Arnold Krammer, "Technological Transfer as War Booty: The U.S. Technical Oil Mission to Europe, 1945," *Technology and Culture* 22 (1981):68–103.

20. Borkin, pp. 157–63; G. D. Taylor, "The Rise and Fall of Antitrust in Occupied Germany, 1945–48," *Prologue*, Spring, 1979, pp. 23–39.

CHAPTER THIRTEEN

1. On postwar developments in the chemical industry, see "The Chemical Century," *Fortune*, Mar., 1950, pp. 69–76, 114–22. On Dow, see Don Whitehead, *The Dow Story* (New York: McGraw-Hill, 1968); and Stocking and Watkins, pp. 274–302.

2. On Monsanto and Chemstrand, see Dan J. Forrestal, *The Monsanto Story: Faith, Hope, and $5000* (New York: Simon and Schuster, 1977), pp. 121–29.

3. *Business Week*, Apr. 20, 1957, pp. 89–93; Perrin Stryker, "Chemicals: The Ball is Over," *Fortune*, Oct., 1961, pp. 125–27, 207–18.

4. Theodore Kovaleff, *Business and Government During the Eisenhower Administration: A Study of the Antitrust Division of the Justice Department* (Athens: Ohio University Press, 1980), pp. 91–93; J. B. Dirlam and I. M. Stelzer, "The Cellophane Labyrinth," *Antitrust Bulletin* 1, no. 9 (1956):635–36.

5. Kovaleff, "Divorce American Style," pp. 29–30.

6. Walter Carpenter, Jr., to Walter Beadle, Treasurer, July 15, 1946; Carpenter, Memorandum on G. M. Stock, Jan. 21, 1947; Pierre du Pont to Carpenter, Jan. 23, 1947 (Carpenter MS, box 833, file 21).

7. Kovaleff, "Divorce American Style," pp. 31–36; Mueller, "Du Pont," pp. 130–40.

8. Kovaleff, "Divorce American Style," p. 35; Zilg, pp. 386–87.

9. *U.S. v. E. I. du Pont de Nemours & Co., Inc. et al.* 126 *Federal Supplement* 235 (1954).

10. Ibid., 353 *U.S.* 586–607 (1957); *Business Week*, June 8, 1957, pp. 41–43. Zilg, p. 390, suggests that the Court's decision may have been influenced by a Du Pont announcement in 1955 that it planned to invest an additional $75 million in G.M., "a direct slap in the government's face."

11. Carpenter, Memorandum for Finance Committee, Oct. 15, 1956; L. Schreiber, "Tax Consequences of Disposition of General Motors Investment," Oct. 23, 1956 (Carpenter MS, box 852, files CO 13.2, CO 3.2); *Business Week*, Oct. 5, 1957, p. 50; Kovaleff, "Divorce American Style," pp. 38–39; 177 *Federal Supplement* 1, 42, 51 (1959).

12. *U.S. v. E. I. du Pont de Nemours & Co., Inc., et al.* 366 *U.S.* 316, 331–32 (1961).

13. Kovaleff, "Divorce American Style," pp. 41–42.

14. *Business Week*, June 8, 1957, p. 43; Cray, *Chrome Colossus*, pp. 386–87.

15. W. R. Swint, Foreign Relations Dept., to J. W. McCoy, Vice-President, July 11, 1945; John K. Jenney, Memorandum on Foreign Activities, Dec. 9, 1936; Jenney to J. E. Crane, July 20, 1937; "Analysis of Origins of Du Pont Sales," 1944 (Crane MS, box 1039).

16. Swint to McCoy, July 11, 1945; Swint to Carpenter, "Du Pont Policy on South American Investments," Nov. 24, 1942; Swint to Carpenter, Dec. 9, 1946 (Carpenter MS, box 832, file 20).

17. Jenney, interview with author, Sept. 2, 1981.

18. Jenney, Columbia interview; Swint to Carpenter, Jan. 4, 1944; Swint to Carpenter, "Memorandum of Conversation with Baron Boel of Solvay," Feb. 23, 1944 (Carpenter MS, box 832); William Richter, Sales Manager, Fabrics and Finishes Dept., to Foreign Relations Dept., Jan. 23, 1933: "Societe Francaise Duco"; Richter to F. W. Pickard, Oct. 1, 1935 (Crane MS, box 1039).

19. Carpenter, "Remarks Concerning Proposed Investment in the Heavy Chemicals Industry in Brazil," Nov. 8, 1942; Swint to Carpenter, Dec. 9, 1946 (Carpenter MS, box 832, file 20); Jenney, interview with author.

20. Carpenter to Swint, Nov. 19, 1946; Swint to Carpenter, Dec. 9, 1946 (Carpenter MS, box 832, file 20).

21. Jenney, Columbia interview; J. B. Eliason, Treasurer's Office, to J. E. Crane, "Return on Investment Requirements for Certain Foreign Countries," Jan. 2, 1946 (Crane MS, box 1039).

22. See Wilkins, pp. 328–34; and Burton Kaufman, *Trade and Aid: Eisenhower's Foreign Economic Policy* (Baltimore: Johns Hopkins University Press, 1982), pp. 152–66, on the foreign aid and investment policies of the U.S. government in this period.

23. Jenney, Columbia interview; David H. Conklin, interview with author, Aug. 23, 1982.

24. *Business Week*, Aug. 15, 1959, p. 97; Conklin, interview with author.

25. Du Pont Tariff Division, "The Du Pont Company and Tariffs," May, 1964 (MS at Eleutherian Mills Historical Library). On the Kennedy Round and chemical tariffs, see Ernest H. Preeg, *Traders and Diplomats* (Washington, D.C.: Brookings Institution, 1970), pp. 109–10, 190–92, 204–5; John W. Evans, *The Kennedy Round and American Trade Policy* (Cambridge: Harvard University Press, 1971), pp. 227–29, 272–73, 285–86, 301–5.

26. Quoted in Zilg, p. 410.

27. Conklin, interview with author; Dmitri Andriadis, interview with author, Aug. 23, 1982.

28. See M. Z. Brooke and H. L. Remmers, *The Strategy of Multinational Enterprise: Organization and Finance* (London: Pitman, 1978), pp. 21–36.

29. Charles B. McCoy, "The Outlook for Du Pont," Remarks before the New York Society of Security Analysts, Oct. 4, 1972; Du Pont, *Annual Report* (1978), p. 26.

30. Lawrence G. Franko, *The European Multinationals* (New York: Harper & Row, 1976), pp. 99–100, 165;–66; Gilbert Burck, "Europe's Industries—Chemicals: The Reluctant Competitors," *Fortune*, Nov., 1963, pp. 148–53, 218–19.

CHAPTER FOURTEEN

1. Figures are based on information compiled from Du Pont's *Annual Reports* for the period 1964–97.

2. McCoy, "The Outlook for Du Pont: 'What's Doing At Du Pont?' " *Forbes*, Nov. 1, 1976, p. 34.

3. W. D. Lawson, "History and Analysis of Corfam," Nov., 1972 (MS at Eleutherian Mills Historical Library).

4. "New Glut of Chemicals in Europe," *Business Week*, Mar. 17, 1973, p. 78; "New World for Chemical Giants," ibid., Sept. 7, 1974, pp. 52–53; "How Bayer is Building the U.S. Market," ibid., Aug. 1, 1977, pp. 72–73; Paul Gibson, "How the Germans Dominate the World Chemical Industry," *Forbes*, Oct. 13, 1980, pp. 155–64; Mario Müller, "Germany's Big Three Find U.S. Pastures Verdant," *German Tribune*, Nov. 16, 1980, p. 7.

5. *Chemical Week*, Feb. 23, 1977, pp. 58, 63; ibid., Mar. 27, 1977, pp. 52–61; ibid., Aug. 10, 1977, pp. 46–48; Carl Gerstacker, "Chemistry in the 1980s."

6. Du Pont, *Annual Report*, 1979, p. 24.

7. *Business Week*, July 21, 1973, pp. 22–23; "Shapiro Takes Over," *Fortune* Jan., 1974, pp. 78–81; Walter Carpenter, Jr., to Lammot du Pont, June 7, 1949 (Carpenter MS, box 849, file CO-7).

8. Geoffrey Smith, "Du Pont Shines Again," *Forbes*, Mar. 5, 1979, pp. 35–36; Lee Smith, "Dow versus Du Pont: Rival Formulas for Leadership," *Fortune*, Sept. 10, 1979, pp. 74–84; "Du Pont–Conoco: Making the Marriage Work," *Chemical Week*, Sept. 2, 1981, p. 50. On the Business Roundtable, see Walter Guzzardi, Jr., "Business is Learning How to Win in Washington," *Fortune*, Mar. 27, 1978, pp. 53–58; Kim McQuaid, *Big Business and Presidential Power* (New York: William Morrow, 1982), pp. 284–95.

BIBLIOGRAPHICAL ESSAY

The literature on the history of the chemical industry is substantial and impressive. Histories of the Du Pont family and firm are equally prolific but vary considerably in quality. The same statement can be made about the historiography of other major national and international chemical enterprises, with the significant exception of the British chemical industry, where there have been some major contributions in recent years.

To a significant extent this study of Du Pont and the international chemical industry is based on primary sources, including the archival holdings of the Eleutherian Mills Historical Library in Greenville, Delaware, and numerous government materials, particularly court records, which have the disadvantage of focusing on aspects of Du Pont's operations that are not always pertinent to the main themes of the company's development. Furthermore, the Du Pont records at Eleutherian Mills are by no means complete, particularly for the period since the 1920s. For various reasons the company has been reluctant to allow outsiders access to such crucial materials as the minutes of the board of directors, and the Executive and Finance committees, and it has allowed valuable records to be destroyed. Useful interviews were conducted with retired Du Pont managers, but some of those approached, notably one former chief executive of the company, declined to be interviewed. Consequently this book cannot claim to provide a definitive account of Du Pont's history, although we have endeavored to present the most complete study possible on the basis of materials available to people outside the corporation and the Du Pont family.

1. The Chemical Industry

Two studies of the chemical industry are of inestimable value to historians. L. F. Haber's *The Chemical Industry During the Nineteenth Century*, 2d ed. (Oxford: Clarendon Press, 1969) and *The Chemical Industry, 1900–1930* (Oxford: Clarendon Press, 1971) constitute the most thorough general surveys of the development of the international chemical industry now available, integrating analysis of technical and commercial aspects in magisterial fashion. Williams Haynes's six-volume history, *The American Chemical Industry* (New York: Van Nostrand, 1945–54), provides exhaustive detail on developments in the United States, particularly on the crucial period 1914–28 which is covered in volumes 2–4. To some extent, Haynes's study reflects the particular biases of American chemical manufacturers, who commissioned the history, especially on the subject of the development of the domestic dye industry; but the author's comprehensive coverage of the field is remarkable. Volume 6 is particularly valuable, since it includes a historical compendium of virtually every firm operating in the industry in the United States.

Several other books provide useful insights on special aspects of the industry. Frank Sherwood Taylor, *A History of Industrial Chemistry* (London: Heinemann, 1957), focuses on major technical developments, particularly in the nineteenth and twentieth centuries. J. R. Partington, *A History of Chemistry,* 4 vols. (London: Macmillan, 1961–70), reviews the development of the science, with minor attention given to industrial and commercial aspects. Paul Hohenberg, *Chemicals in Western Europe, 1854–1914: An Economic History of Technical Change* (Chicago: Rand-McNally, 1967), provides an overview of the growth of the industry in Europe.

2. Du Pont

Du Pont has been the subject of numerous books and articles. Unhappily, few of them are equal in quality and comprehensiveness to the more general works of Haber and Haynes, which do contain some valuable insights on this most important chemical company. There are a variety of popular accounts of the Du Pont family, most of which concentrate on the early years of the company and such melodramatic events as the struggle for control of the firm during World War I. Recent examples of this literature include Leonard Mosley, *Blood Relations: The Rise and Fall of the Du Ponts* (New York: Atheneum, 1980); John D. Gates, *The Du Pont Family* (Garden City: Doubleday, 1979); William H. Carr, *The Du Ponts of Delaware* (New York: Dodd Mead & Co. 1964); and Max Dorion, *The Du Ponts: From Gunpowder to Nylon* (Boston: Little, Brown, 1962). Marquis James, *Alfred I. du Pont: The Family Rebel* (New York: Bobbs-Merrill, 1941), provides details on the family power struggle from the viewpoint of one of its main protangonists. Few of these works devote much attention to the organization and functioning of Du Pont as a business enterprise. William S. Dutton, *Du Pont: One Hundred and Forty Years* (New York: Scribners, 1942), provides more detail on this subject. Dutton, as a former employee of the firm, had access to company records but adopts a generally uncritical stance toward his subject.

One of the most detailed, and most controversial accounts of the history of the Du Pont family and firm is Gerald C. Zilg, *Du Pont: Behind the Nylon Curtain* (Englewood Cliffs, N.J.: Prentice-Hall, 1974). Zilg's study is extremely critical of both family and firm and attempts to place his subjects in the perspective of a generally radical critique of industrial capitalism in America, derived in part from the views of historians such as William Appleman Williams and Martin Sklar. Much of Zilg's work is based on published sources as he was, by his own account, "barred from all post 1933 manuscripts"—and evidently used relatively little of the primary source material before that date as well. The analysis tends to be somewhat simplistic on organizational aspects of the development of Du Pont, and the author has a tendency to chastise Du Pont for the sins of capitalism in general, even where such criticism may not be appropriate. Still, Zilg does attempt to discuss Du Pont's growth in a broader framework than most other popular accounts.

Another critical study that seeks to introduce a more sophisticated approach is Stephen J. McNamee, "Du Pont–State Linkages: A Socio-Historical Analysis" (Ph.D. dissertation, University of Illinois, 1980). McNamee's argument overstates the extent of Du Pont's connections with the U.S. government, implying a continuity and convergence of interest that is not always apparent. Perhaps the most useful of the various critical works on Du Pont is Willard F. Mueller, "Du Pont: A Study in Firm Growth" (Ph.D. dissertation, Vanderbilt University, 1955), that drew extensively on antitrust case records involving Du Pont, and reflects generally the viewpoint of the Justice Department of the company as a habitual offender against the Sherman and Clayton Acts. (Mueller's thesis adviser was George Stocking, a consultant to the Antitrust Division in the 1940s and 1950s.) Some of the most interesting elements of this account deal with Du Pont's strategy of technical development, an area expanded upon in Mueller, "The Origins of the Basic Inventions Underlying Du Pont's Major Product and Process Innovations, 1920 to 1950," in National Bureau of Economic Research, *The Rate and Direction of Inventive Activity: Economic and Social Factors* (Princeton: N.B.E.R., 1962), pp. 323–46.

Alfred D. Chandler, Jr., and Stephen Salsbury, *Pierre D. du Pont and the Making of the Modern Corporation* (New York: Harper & Row, 1971), is unquestionably the best study of any aspect of Du Pont. Based on extensive research in company archives (much of it, unfortunately, not likely to be made available to future scholars), this book reviews the business career of one of the major figures in the development of both Du Pont and General Motors, and provides a detailed analysis of the technical and organizational changes at Du Pont at a critical point in the company's history. This study complements chapters 2 and 3 of Chandler's *Strategy and Structure* (Boston: M.I.T. Press, 1962), which traces the evolution of decentralization at Du Pont in 1919–21, and its influence on the reorganization of G.M. in this same period.

Other published materials that provide insight on various aspects of Du Pont include A. P. Van Gelder and Hugo Schlatter, *History of the Explosives Industry in America* (New York: Columbia University Press 1927), which discusses the development of the company and the industry in the nineteenth century; N. B. Wilkinson, "In Anticipation of Frederick Taylor: A Study of Work by Lammot du Pont," *Technology and Culture* 6 (Spring, 1965):208–21; and Ernest Dale and Charles Meloy, "Hamilton M. Barksdale and the Du Pont Contributions to Systematic Management," *Business History Review* 36 (Summer, 1962):127–52, deal with organizational innovations at Du Pont in the late nineteenth and early twentieth centuries. E. K. Bolton, "Du Pont Research," Address to the Society of the Chemical Industry, Jan., 1945 (reprinted in Du Pont, *Annual Report* [1945]), and Lawrence Lessing, "Du Pont: How to Win at Research," *Fortune*, Oct., 1950, pp. 115–18, 122–34, focus on this important area of Du Pont's operations and should be complemented by Mueller's work cited earlier. Theodore Kovaleff, "Divorce American Style: the Du Pont–General Motors Case," *Delaware History* 18, no. 1 (1978):28–42, reviews this case that had dramatic effects on the company, and also traces the development of the Du Pont–G.M. relationship. A more pedestrian narrative is in Arthur Welsh, "The Du Pont General Motors Case" (Ph.D. dissertation, University of Illinois, 1963).

The most important primary source materials used in this book are in the Records of the E. I. du Pont de Nemours & Company, Inc., ser. 2, pt. 2, at Eleutherian Mills Historical Library. These include the papers of Pierre S. du Pont and Walter S. Carpenter, Jr.; a body of papers designated "Administrative Files" that contain papers of Irenee du Pont and Lammot du Pont during their terms as chief executives of the company between 1919 and 1940; the papers of Jasper E. Crane, a vice-president of Du Pont from 1929 to 1945 who was responsible for the company's "foreign relations" during this period; records of the Foreign Relations Department; and records of the Legal Department, including trial transcripts and exhibits for a number of cases in which Du Pont was involved. The most important case records include: *U.S.* v. *E. I. du Pont de Nemours & Co., Inc., Imperial Chemical Industries Ltd., et al.* (Civil Action 24–13, Southern District of New York, 1944); and *U.S.* v. *E. I. du Pont de Nemours & Co. Inc., General Motors Corporation, et al.* (Civil Action 49 C–1071, Northern District of Illinois, 1949). The records of these cases held by the U.S. Justice Department in Washington, D.C., were also reviewed.

In addition, we drew upon Du Pont's *Annual Reports*, and on interviews with retired Du Pont employees, including John Jenney, who was involved with the company's Foreign Relations Department from 1927 to 1967; David H. Conklin, who headed Du Pont's British and European operations from 1957 to 1973; and Richard Manning, who was general counsel in the Legal Department from 1955 to 1970, and was involved in the latter stages of the I.C.I. case. Work with primary sources and the secondary sources noted above was supplemented by extensive use of published material in commercial and technical journals cited in the notes, including *Chemical and Engineering News, Chemical Week, Business Week, Forbes, Fortune,* and the *Wall Street Journal*.

Records and publications of the United States government also proved to be useful, although some material, particularly in congressional hearings, must be approached with caution, keeping in mind the political circumstances underlying them. Among the most pertinent of these congressional materials are U.S. Congress, House, Ways and Means Committee, *Hearings on H. R. 2706 and H. R. 6495: Dyestuffs* (Washington, D.C.: G.P.O., 1919); and U.S. Congress, Senate, Judiciary Committee, *Hearings on Alleged Dye Monopoly* (Washington, D.C.: G.P.O., 1922), which deal with the Chemical Foundation and the dye tariff legislation in 1919–22; U.S. Congress, Special Committee to Investigate the Munitions Industry, *Hearings* (Washington, D.C.: G.P.O., 1936), the "Nye Committee"; U.S. Congress, Senate, Temporary National Economic Committee, *Investigation of Concentration of Economic Power: Hearings* and *Monographs* (Washington, D.C.: G.P.O., 1939–41), which includes material on patents and trade association activities; and U.S. Congress, Senate, Committee on Military Affairs,

Hearings on Scientific and Technical Mobilization (Washington, D.C.: G.P.O., 1943–44), which includes material relating to international cartels and patent agreements in the 1930s.

Other government sources used include the U.S. Tariff Commission, *Census of Dyes and Synthetic Organic Materials* (Washington, D.C.: G.P.O., 1918–26), which provides details on the development of the dye industry in the United States and abroad; Records and Briefs of *U.S. v. The Chemical Foundation Inc.* 272 *U.S.* 1 (1923) at the Library of Congress in Washington, D.C.; and the reports on various cases involving Du Pont cited in the notes.

3. Other Chemical Companies

Literature on other American chemical firms is rather limited. Beyond the brief histories in Haynes, *American Chemical Industry*, vol. 6, and various articles in business and technical periodicals cited in the notes, there are relatively few substantial works on companies such as Allied Chemical and Dye, American Cyanamid, Union Carbide, and General Aniline. Don Whitehead, *The Dow Story* (New York: McGraw-Hill, 1968), and Dan J. Forrestal, *The Monsanto Story: Faith, Hope and $5000* (New York: Simon and Schuster, 1977), retail the histories of these two firms and are based on company archival sources, but share the same defects as many of the popular histories of Du Pont in that they are not well documented and suffer a lack of perspective on their subjects. A notable exception to this rule is Edith T. Penrose, "The Growth of the Firm—A Case Study: The Hercules Powder Company," *Business History Review* 34 (Winter, 1960):1–23, whose findings were integrated into her major theoretical work, *The Theory of the Growth of the Firm* (Oxford: Blackwell, 1959). An insightful analysis of one branch of the American chemical industry is Martha M. Trescott, *The Rise of the American Electrochemical Industry, 1850–1910* (Westport, Conn.: Greenwood Press, 1981).

For histories of major chemical companies outside the United States, the British have been best served, thanks to a tradition of sophisticated and professional company histories, and a relatively enlightened attitude on the part of the companies toward those who would review their past. A good general survey of the industry is in D. W. F. Hardie and J. Davidson Pratt, *A History of the Modern British Chemical Industry* (Oxford: Pergamon Press, 1966). Of particular value to the study of Du Pont is William J. Reader's two volume *Imperial Chemical Industries: A History* (London: Oxford University Press, 1970–75), which moves far beyond its immediate subject to survey the network of relations among international firms comprising the chemical industry. Also of value is Donald C. Coleman's three-volume work, *Courtaulds: An Economic and Social History* (Oxford: Clarendon Press, 1969–1980), which reviews the development of this pioneer in the rayon field; and Charles Wilson, *The History of Unilever* (London: Cassell, 1954) in two volumes.

The Swiss chemical and pharmaceutical firms have yet to be assessed in a good general history. There is a history of Geigy, A. Burgin, *Geschichte des Geigy Unter-*

nehmens von 1758 bis 1939 (Basle: Birkhauser, 1958), and a brief historical introduction in Paul Erni, *The Basel Marriage: A History of the Ciba-Geigy Merger* (Zurich: Neue Zürcher Zeitung, 1979). Solvay of Belgium receives attention in Reader's first volume on I.C.I., and is also memorialized in Jacques Bolle, *Solvay: L'Invention, L'homme, L'enterprise industrielle* (Brussels: Weissenbruch, 1963). A recent study of the development of chemicals in France in the early nineteenth century is John Graham Smith, *The Origins and Early Development of the Heavy Chemical Industry in France* (New York: Oxford University Press, 1980).

Paradoxically the German chemical industry, dominant in a range of fields and a major element in world chemical markets over the past one hundred and fifty years, has yet to be chronicled on a level equal to that of the British and American industries. Haber discusses the growth of the German industry to 1930 in his two volumes, and John J. Beer, *The Emergence of the German Dye Industry* (Urbana: University of Illinois Press, 1959) reviews that central area of the industry up to World War I.

I. G. Farben, by Haber's reckoning, was the subject of more than four hundred articles and two novels in Germany, but no definitive history of that company has yet been published. Fritz ter Meer's *Die I. G. Farben* (Dusseldorf: Econ, 1953) is an apologia written by a former director of the firm who was prosecuted at Nuremberg. There are various demonologies prepared by American lawyers who were involved in phases of the prosecution of I. G. Farben after World War II, including Richard Sasuly, *I. G. Farben* (New York: Boni and Gaer, 1947); Josiah E. Dubois, *The Devil's Chemists* (Boston: Beacon Press, 1947); and Joseph Borkin, *The Crime and Punishment of I. G. Farben* (New York: Macmillan, 1978), which is the best of this group. All of these accounts focus on I.G.'s involvement in Nazi war crimes and related subjects. The prosecution of I. G. Farben at Nuremberg (headed by Dubois) prepared a two-volume history, "Basic Information on I. G. Farbenindustrie," based on documents seized at Frankfurt in 1945. Other accounts in German relating to I. G. Farben and its member firms include Carl Duisberg, *Meine Lebenserinnerungen* (Leipzig: P. Reclam, 1933); H. J. Flechtner, *Carl Duisberg* (Dusseldorf: Econ, 1960); F. Jacobi, ed., *Beiträge zur hundertjahrigen Firmengeschichte* (Dusseldorf: Econ, 1964), principally on Bayer; O. Steinert and W. Roggersdorf, *Im Reiche der Chemie—100 Jahre BASF* (Dusseldorf: Econ, 1965); and E. Bäumler, *Ein Jahrshundert Chemie* (Dusseldorf: Econ, 1963), on Hoechst.

There are also specialized studies of I. G. Farben's involvement with synthetic fuels and synthetic rubber, including Wolfgang Birkenfeld, *Der Synthetische Treibstoff, 1933–1945* (Gottingen: Musterschmidt, 1964); Thomas P. Hughes, "Technological Momentum in History: Hydrogenation in Germany, 1898–1933," *Past and Present* 44 (Aug., 1969):106–32; and Arnold Krammer, "Fueling the Third Reich," *Technology and Culture* 19 (1978):394–422.

The literature on international cartels is enormous and much of it focuses on the chemical industry. Most pertinent are chapters 3 and 9–11 in Reader's second volume on I.C.I.; G. W. Stocking and M. W. Watkins, *Cartels in Action: Case Studies in International Business Diplomacy* (New York: Twentieth Century Fund, 1946); Kurt

R. Mirow and Harry Mauser, *Webs of Power: International Cartels and the World Economy* (Boston: Houghton Mifflin, 1982); Ervin Hexner, *International Cartels* (Chapel Hill: University of North Carolina Press, 1946); Gabriel Kolko, "American Business and Germany, 1930 to 1941," *Western Political Quarterly* 15 (Dec., 1962): 713–28; and Mark Lytle, "Thurman Arnold and the Wartime Cartels" (manuscript, Sterling Memorial Library, Yale University). Mira Wilkins in *The Maturing of Multinational Enterprise* (Cambridge: Harvard University Press, 1974), pp. 78–82, 263–67, offers a different vantage point toward these cartels from the other sources that tend to reflect the position of American antitrust authorities. I. G. Farben's connections with Standard Oil (New Jersey) are discussed in H. M. Larson, E. H. Knowlton, and C. S. Popple, *History of Standard Oil Company (New Jersey): New Horizons, 1927–1950* (New York: Harper & Row 1971), pp. 153–59, 405–18.

Du Pont's relations with I.G. in the 1920s and 1930s, from the perspective of antitrust officials, are summarized in a 1943 memorandum, "Du Pont–I.G. Relationship," in the papers of Roy Prewitt, Federal Trade Commission Records, Record Group 122, U.S. National Archives, Washington, D.C. There are two other collections of manuscripts in the National Archives that are particularly valuable on I. G. Farben and its international operations. In the Collection of Seized Enemy Records, Record Group 242, are photocopied materials on I. G. Farben, including summaries of all its formal agreements with foreign companies. In the Records of the Office of the Alien Property Custodian, Record Group 131, is a large collection of materials seized from Chemnyco Inc., a subsidiary of I.G. in the United States, which contain voluminous material on the entire chemical industry between 1929 and 1941.

INDEX

ABOUT THE AUTHORS

Graham D. Taylor is an associate professor of history at Dalhousie University in Halifax, Nova Scotia. He received his Ph.D. at the University of Pennsylvania in 1972. He was a coeditor of *The New American State Papers* (1973–74), author of *The New Deal and American Indian Tribalism* (1980), and has published a number of articles on business and diplomatic history in such journals as *Prologue*, the *Business History Review, Diplomatic History*, and the *International History Review.*

Patricia E. Sudnik is completing a doctoral thesis at the University of Chicago on American foreign policy and multinational enterprises between 1920 and 1940. She teaches in the history department at Seton Hill College in Greensburg, Pennsylvania.